SEX ON THE KITCHEN TABLE

The Heart of the Love Apple, by Beverly Ellstrand. This painting depicts the results of plant romance; a cutaway crosswise view, the fruit and seeds of the tomato, aka the aphrodisiacal *pomme d'amour*.

Sex on the Kitchen Table

THE ROMANCE OF PLANTS AND YOUR FOOD

Norman C. Ellstrand

ILLUSTRATED BY SYLVIA M. HEREDIA

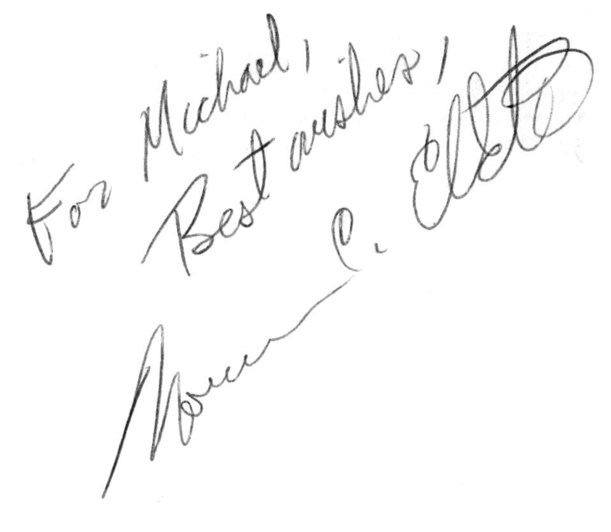

The University of Chicago Press
Chicago and London

The University of Chicago Press, Chicago 60637
The University of Chicago Press, Ltd., London

Published 2018
Printed in the United States of America

27 26 25 24 23 22 21 20 19   2 3 4 5

ISBN-13: 978-0-226-57475-2 (cloth)
ISBN-13: 978-0-226-57489-9 (paper)
ISBN-13: 978-0-226-57492-9 (e-book)
DOI: https://doi.org/10.7208/chicago/9780226574929.001.0001

Library of Congress Cataloging-in-Publication Data

Names: Ellstrand, Norman Carl, author. | Heredia, Sylvia M., illustrator.
Title: Sex on the kitchen table: the romance of plants and your food / Norman C. Ellstrand; illustrated by Sylvia M. Heredia.
Description: Chicago; London: The University of Chicago Press, 2018. | Includes bibliographical references and index.
Identifiers: LCCN 2018006588 | ISBN 9780226574752 (cloth: alk. paper) | ISBN 9780226574899 (pbk: alk. paper) | ISBN 9780226574929 (e-book)
Subjects: LCSH: Plants—Reproduction. | Vegetable gardening. | Cooking, American.
Classification: LCC QK825 .E55 2018 | DDC 581.3—dc23
LC record available at https://lccn.loc.gov/2018006588

∞ This paper meets the requirements of ANSI/NISO Z39.48-1992 (Permanence of Paper).

For Tracy, Nathan, Mom, and Dad

Contents

Preface

> . . . [it's] amazing that the human race has taken enough time out from thinking about food or sex to create the arts and sciences.
> —Mason Cooley, professor, writer, and aphorist

Amazing, indeed! There's no question that we spend a lot of time thinking about food or sex. But do we think more about food or more about sex? Today my Google search for the word "sex" generated 3.2 billion results, while "food" totaled almost 4.2 billion. It makes sense for food to be number one. After all, food is essential for survival. Sex is often nice but not a necessity. In fact, food and sex have a natural affinity. Are we surprised to find that a Google search for both "food" *and* "sex" yields almost a half billion hits? The combination of those elements became the alloy that built my scientific career.

My path to studying the intersection of sex, food, and science wasn't entirely straightforward. I was raised by 1960s foodies, even if the term had not yet been coined. With one parent of eastern European descent and the other of Scandinavian parentage, celebrating interesting and diverse food was part of the program. Pickled herring and lox were mentioned more frequently than fish sticks. The virtues of bagels and chewy rye bread were celebrated, but not those of vitamin-fortified white bread. Restaurant visits were joyful experiments: "Norm, why don't *you* try the escargot?"

Addictive bird-watching set me up to be a sap for evolutionary biology in college, while my interest in sex had a more organic basis. My undergraduate mentor, University of Illinois professor David Nanney, nourished the curiosity, asking me an innocent couplet of questions: "Can reproduction occur without sex? Can sex occur without reproduction?" I started to read up on the topic. Imagine my happy surprise to learn that evolutionary scholars of the mid-1970s were engaging in a challenging theoretical debate about why sex should occur at all.

Why sex? I was hooked on the topic and ready to take it up in grad school. The transition from birds to plants was easy. Professor Don Levin sold plants as ideal study organisms: "They don't move, bite, bleed, or poop in your hand. . . . Zoologists come up with the good ideas, but their organisms are lousy." Furthermore, plants offer a selection of sexual diversity that makes bird sex seem vanilla by contrast. Of course, plant science includes agriculture—and the biology of food plants. Even though I studied wild species for my degree, I learned that much of the best work on plant evolution was pioneered by scientists who worked on crop species. One of those pioneers was Professor Janis Antonovics at Duke University. Just coming back to the United States after a year of figuring out how to boost Taiwan's rice yields, he designed the first set of experiments to get at why sex is so common throughout all the kingdoms of life. He hired me as a postdoctoral researcher. Antonovics proved a valuable mentor, teaching me how to work smart instead of merely working hard. His advice: "Do what gives you energy."

What energized me was figuring out how to stir sex and food together. The University of California, Riverside, is a land-grant university, so my experiment station funding was intended to benefit California agriculture. Food plants at last! I maintained both basic and applied research programs for about a decade. The two fused into one when I realized that plant sex could deliver engineered genes to unintended populations—in particular, those of wild crop relatives. Bringing the science of plant sex to inform GMO crop regulatory policy has been a decades-long adventure that hasn't yet ended. Of the many things I've learned from the adventure is that the role of sex in producing food is largely misunderstood, even among many scientists. When I entered GMO crop policy discussions, I was soon humbled by my lack of knowledge about agronomy and horticulture and how little I knew about bringing engineered plants from cell culture to market. Along the way, I educated other scientists in population genetics and evolutionary biology. I learned that the phrase "Everybody knows that . . ." often ends in a falsehood, independent of the source's lips. That's one motivation for this foodie sex manual.

Humans have a necessary relationship with food. The deluge of food-related information into our lives—from authentic to purely fantasy to intentional fake news—interferes with building that relationship. Romance with what we eat should be lasting; it cannot be based on either *plat du jour* infatuation or anxiety-stuffed YouTube videos. In recent years I've witnessed a blog war of half-baked fire-and-brimstone self-ordained food preachers ("Would you like some whine with that wafer?") versus smug, self-righteous pseudo-scientists ("How could they be so dumb? Everybody knows that . . ."). Don't let the charlatan behind the curtain bamboozle you. (I myself went off the tracks at least once. In retrospect, a character-building experience that I don't want to repeat.)

We need to get to know our food. This book is a vehicle toward that goal. Let each chapter about food and sex be a tête-à-tête, the slow development of a lasting romance through understanding what we love and what sustains us. Scientific understanding should be accessible and fun. Go beyond this book. Question everything. Don't let what you believe interfere with gaining knowledge. And remember that scientific information is constantly changing and improving. Keep learning. Use what you learn here and elsewhere to make your own ethical, environmental, political, and social opinions about what you like and what you don't like. Then act on your now scientifically informed but always open to new information opinions.

Dim the lights; turn on the mood music; proceed to chapter 1.

Bon appétit!

1 *The Garden*

> And out of the ground made the Lord God to grow every tree that is pleasant to the sight, and good for food; the tree of Life, also in the middle of the Garden, and the tree of Knowledge of Good and Evil.
> —Genesis 2:9 (American King James Version)

There's lots of drama with the Bible's first mention of plants intentionally grown for food—Life, Birth, and Death; Knowledge, Innocence, and Ignorance. In the next chapter of Genesis, the Serpent's dietary advice to Eve starts a chain of events that reveals the intimate relationship between food and sex. Sustainable self-nourishment, life, depends on the reproduction, the birth, and the death of other living organisms. Food depends on sex. Sex depends on food.

A question we ask as children is "Where did I come from?" Here we explore another question about sex: "Where does our food come from?" Like Eve and Adam, most of us feel estranged from our Garden. No surprise. In the early years of the twenty-first century, humanity entered an era in which more than half of humanity became urbanites (United Nations 2014). Only one in five individuals is directly involved in agriculture (Alston and Pardey 2014). No wonder urbanites hunger for knowledge about food, stimulating an

FIGURE 1.1 In the Garden. The Serpent suggests fruit for the dessert course.

ever-increasing flood of printed and electronic attention to the subject.

In particular, the last decade has seen a surge in interest in all things food. At last, after years in the closet, food is now a legitimate research area for scholars beyond agricultural scientists and food processing engineers. Likewise, food has come to center stage in popular culture. Witness the Food Network and the proliferation of foodie movies: *Chef*, *Basmati Blues*, *Ratatouille*, *The Hundred-Foot Journey*, *Chocolat*. Food personalities not only include chefs, celebrity and otherwise: Alice Waters, Thomas Keller, Wolfgang Puck, Jamie Oliver, Martha Stewart, Ferran Adrià, Gordon Ramsay, and Rick Bayless; but also savvy observers who go beyond restaurant and recipe: Tony Bourdain, Dan Charles, and Marion Nestle. A few food pundits have noted that the media's celebration of food has parallels with sex. The sensuality of their visuals has been characterized as "food porn."

As interesting and entertaining as this festival of food may be, it still fails to answer the question of where our food ultimately comes from. The source of food can be a profound mystery for those estranged from agriculture. Even well-educated people may be mystified or misinformed. The young man next to me on a transatlantic flight complained that if we could just get rid of genetically engineered crops, we could go back to eating wild plants. His jaw dropped when I explained that for the vast majority of humanity, wild plants had dropped off the global menu thousands of years ago. This MBA from a major public university had never learned the simple fact that non-engineered domesticated food plants are profoundly genetically different from wild plants, that the plant and animal foodstuffs he eats are products of thousands of years of evolution directed by human hands. As we shall see, food plants, with few exceptions, are the result of both intentional and unintentional genetic modification. The vast majority of these genetic improvements requires sex, which generates variation, and then human selection, which sorts among the variants. Genetic modification? For sure. Genetic engineering? At the moment, for only a handful of species.

The goal of this book is to engage with what goes on in the Garden, to understand the romance of the plants that feed us. *Sex on the Kitchen Table* presents the plant sex behind our food (in its various manifestations of reproduction, evolution, and genetics) as the avenue for understanding where our food comes from in the short term—the tomato in your hand—and the long term: how a scruffy coastal weed evolved into an important source of sugar. Thus, "Where does our food come from?" has a pair of answers:

The first. Much of our plant-based food is the immediate product of sex or closely associated with sex. Seeds roasted for coffee, pulverized for mustard, ground into flour, germinated for beer, and fermented to make cocoa are baby plants, the products of sexual reproduction. Fruits are seed receptacles. Avocados, blueberries, cucumbers, durians, eggplants, feijoas, grapes, hawthorn berries, ilamas, jackfruit, kiwifruit, limes, mangoes, nectarines, olives, peppers, quinces, rambutans, sweetsops, tomatoes, *umbú*, vanilla "beans," watermelons, *xoconostle*, yellow sapotes, and zucchini are fruits whose function is to hold the seeds created by sexual reproduction. Certain floral organs, so obviously sexual—from squash blossoms to saffron—make it into our mouths as well.

The second. Sex plays an arm's-length role in the ultimate origins of the plants we eat. Most food plants are the result of hundreds of generations and thousands of years of evolution, first under human disturbance, followed by human management, then domestication and, eventually, continued genetic improvement. Humans cultivated plants long before they intentionally selected those with superior heritable characteristics. That is, late Stone Age proto-farmers managed the plants they liked and used. First they foraged. Then they began to experiment. Maybe they pruned the branches of fruit trees to stimulate new growth. Or they helped out plants with edible underground parts by digging out nearby competitors. Or maybe they pulled up some plants and replanted them closer to their hut, where the transplants enjoyed nutrients seeping from rotting garbage and other human waste. These early methods of cultivation ended up selecting for genotypes that benefited from manipulation, starting the long process of domestication. Plants nurtured by human care created more seeds and thereby passed on more of their genes. As proto-crops accumulated those inherited traits, they evolved to become more dependent on humans, and human behavior developed to become more dependent on those plants (Pollan 2001). Species that evolved to survive and reproduce under human care are domesticated. Some species are so thoroughly domesticated—such as corn and soy—that they cannot persist more than a generation or two without human intervention (Owen 2005).

Over time, the process of domestication slowly transitioned into a fully intentional process. So-called plant improvement involves generation by generation manipulation of plant lineages in the pursuit of desired characteristics. For example, as I write, wheat breeders throughout the world are scrambling to create new kinds of wheat varieties to meet the challenge of withstanding a new strain of stem rust that first appeared in 1999 in Uganda (Ug99) and has spread to thirteen countries, threatening the food security of Africa (Singh et al. 2011). (Monitor the advance of this disease yourself at http://rusttracker.org.)

I teach a science course on the biology of food to non-majors, that is, students whose majors focus on topics as diverse as art, literature, business, engineering, and theater—anything but natural science. The undergraduates who take my California's Cornucopia class teach me in turn. I have learned that humans, as eating animals, are naturally excited about finding out more about food. Thus I've taken a novel approach. The conventional "useful plants" course marches through the products lecture by lecture: the legumes, the fiber plants, the cereals, et cetera. Instead, I use individual crops as platforms for exploration.

I take the same approach here. This book is designed to be tractable to a broad and curious audience. If you took a biology course too long ago to remember the details, then this book is for you. Maybe you got the general biology down, but your professor shorted you on plant biology. The scientific terminology gradually builds chapter by chapter so that the reader is fully prepared to understand genetic engineering in the last one. Readers well-studied in botany or other plant sciences may want to skim chapter 2. For those readers who do not know plants and genetics intimately, the linear approach is the best.

Chapter 2 is a plant sex manual, using food plants as examples for understanding floral and fruit structure, pollination, and the various spatial and temporal arrangements of maleness and femaleness in different plant species. The tomato, the "love apple," is the standard for comparison. Chapter 3 features the phallic banana as an example of the

FIGURE 1.2 *Sex on the Kitchen Table*'s cornucopia.

perils of life when reproduction occurs without sex. Along the way, we take a side trip to learn about the tension between the short-term economic and evolutionary benefits of uniformity that asexuality begets versus the long-term benefits of diversity created by sex. Getting chapter 4's avocado to your table and into your mouth involves three different roles of reproductive timing: the timing of male and female expression, the year-by-year changes in the number of fruit produced per tree, and the timing of fruit maturity and ripening. The sweet side of human-manipulated plant romance is one of the stories told in chapter 5, whereby increasingly sophisticated plant-mating techniques by human matchmakers molded the evolution of sugarcane's rival, the sugar beet. It also tells a darker erotic tale of how one of those methods unintentionally facilitated a dangerous liaison between the sugar beet and its wild progenitor, leading to the evolution of one of the world's most costly weeds. In chapter 6, squash is the platform for exploring how genetic engineering, a relatively new method of plant breeding, creates new foods by a sexual process that is essentially billions of years old. But sex as we know it can spread those engineered genes around in some surprising ways. The epilogue pulls together what we've learned to examine to the future of our ever-changing relationship with the Garden.

While the scientific prose in these chapters might generate sufficient thought for food, it is no substitute for the empirical experience of enjoying edibles. Therefore, each chapter concludes with the plant or plants of focus being celebrated with a favorite recipe, an opportunity for savoring and reflection.

The point is that we never actually left the Garden.

Garden Soup—aka Gazpacho

Plant scientists are saps for species diversity. We swoon in botanic gardens, and we glow hiking through the wilderness. Farmers' markets make our heads explode. The Garden is the archetype of diversity. There's no better celebration of the Garden than gazpacho because the recipe is so accepting of whatever you've got on hand. My modern version is a distant descendant of a cucumber-vinegar mix for refreshing Roman soldiers on the march.

4 cups tomato and/or vegetable juice
½ cup minced green onion or scallions

1 or 2 medium cloves garlic, crushed
1 medium minced bell pepper
1 teaspoon honey or sugar
Juice of ½ lemon (avoid Meyer lemon, which isn't a real lemon)
Juice of 1 lime (the little seedy ones or the bigger seedless ones are both suitable, but have remarkably different flavors)
1 tablespoon red wine vinegar
1 tablespoon unfiltered apple cider vinegar
1 teaspoon dried basil (or a tablespoon of chopped fresh basil)
1 teaspoon ground cumin
¼ cup minced parsley or cilantro
3 tablespoons olive oil
2 cups seeded and diced fresh ripe tomatoes
Salt, pepper, etc. to taste

Serves six to eight.

Mix the ingredients briefly in a bowl. Reserve a cup or two. Purée the remainder in a blender. Add the reserve to the purée for texture. Chill and serve cold.

Experiment, experiment, experiment! Feel free to add, subtract, and/or substitute to suit your fancy. Just a few examples: add finely chopped fennel or apple, fresh peas fresh from the pod, or maybe cooled cooked shrimp directly to the final chilled soup. Try other seasonings: cayenne, paprika, cardamom . . .

2 Tomato

THE PLANT SEX MANUAL

Love is a many-splendored thing.
—from the 1955 movie and song (lyrics by Paul Francis Weber)

Consider the perfect tomato. Warm from the garden's summer sun, fulsome and smooth, firm but yielding, bursting with flavor, at once tart and sweet. Feel the juice on your chin?

Long beloved and well-domesticated by pre-Columbian civilizations, the tomato immigrated to Europe in the sixteenth century (Rick 1995). But it received a chilly reception on that continent. The plant had striking similarities to well-respected members of the European flora, well-respected for being poisonous. Tomato fruits and flowers bear more than a superficial resemblance to Europe's various nightshades; they are members of the same family. Deadly nightshade is one of the world's classic toxic plants, including its sweet ripe berries. Black nightshade's not-quite-ripe berries contain enough poison to kill a child (Delfelice 2003). Many Europeans reasoned that the New World fruit was obviously poisonous. The tomato's Latin species name, *lycopersicum* (wolf peach), reflects that misconception. Pinch a tomato leaf off

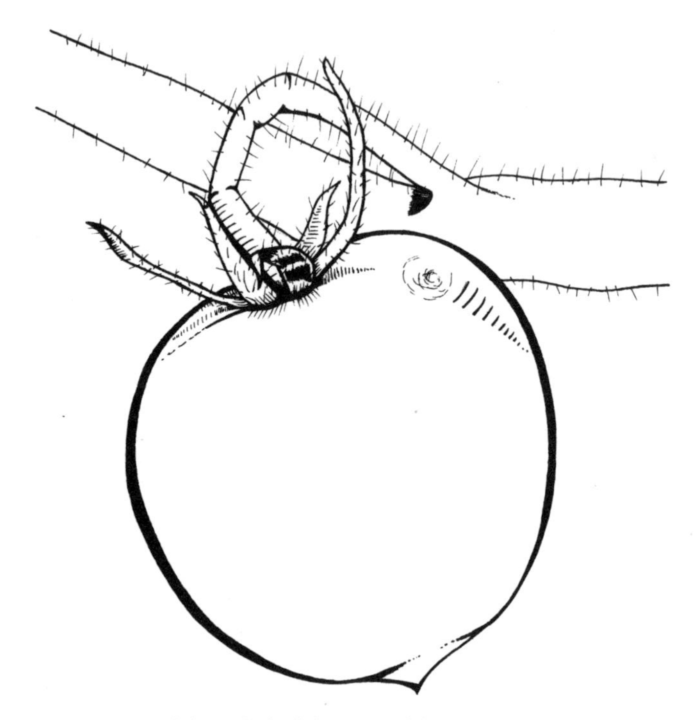

FIGURE 2.1 Tomato fruit attached to hairy stem with five sepals.

a branch, and mash it up in your fingers. Give it a sniff. You know something is a little scary about this plant. Can you blame most Europeans and even European-derived North Americans from keeping the tomato at arm's length as an ornamental for years before wrapping their lips around the ripe fruit?

Ripe tomatoes are red, heart-shaped, and hopelessly sexy. Some romantic individuals believed that the tomato, with proper dosage, had aphrodisiacal properties, which explains its old French name: *la pomme d'amour* (love apple). While the British and their American colonies remained squeamish, Iberians and Italians overcame their initial fears and rapidly incorporated tomatoes into their cuisine. It's hard to think of Italian food without tomatoes. Nonetheless, as a youth, Christopher Columbus ate his pasta without red sauce.

Consider the tomato plant. Looks sort of lonely. But is it?

A tomato plant growing in a field surrounded by its own kind enjoys opportunities for finding sweethearts that rival the singles bar scene. Every plant in a tomato field is sexually receptive to every other. That's not only true for *Solanum lycopersicum*. Tens of thousands of plant species actually live the life of what might be, for some humans, a fantasy. Mutual sexual receptivity is the rule not just for tomato plants but for most plant species (Richards 1997). Indeed, their reality probably exceeds fantasy in terms of the universal availability of mates. In Tomato Town, everybody bears both male and female parts. Everyone is capable of mating with every other individual within pollination distance. In species like the tomato, a plant near ten other tomato plants has eleven potential mates. Eleven? Is that a typo? Nope. Every plant can mate with everyone else *and themselves as well*. The most common form of plant sexuality is the ability to successfully fertilize oneself, what both botanists and zoologists call self-fertility (yep, there are animals that can do this, too). Self-fertile individuals have the potential to serve as both mother and father to one or more of their kids.

As you ponder this situation, consider that you digest the fruits of plants that practice this kind of sexuality regularly. The tomato slice on your burger, the pickle on the side, and the corn kernels processed into high fructose corn syrup in your soda are sexual products of bisexual plants capable of mating with strangers, relatives, or themselves. Most human global nutrition is the direct result of plant sexual activity, seeds and their fruit "packaging." Less frequently, we eat parts of plants one step removed from sexuality: flowers or flower buds. And, of course, we occasionally consume vegetative plant parts—leaves, stems, and roots. From time to time, plants that don't have sex contribute to our diet. Life without sex is far from dull. In fact, it's dangerous. But first let's consider those plants that partake of sex on a regular basis.

While potential promiscuity coupled with self-fertility is the most common mode of plant sexuality, it is far from the only one. The world of plant sex is more multifarious than a menagerie. For plants, love is, indeed, a many-splendored thing, with so much variation it exceeds imagination. That variation extends to the plants that supply us with food. Eros and Aphrodite lurk, giggling, behind the bites that you take. Let's frolic through the sexual diversity with which evolution has blessed plants to fine-tune how their genes move from one generation to the next. The kaleidoscopic array of plant sex behaviors may seem a shocking panoply of practices, but for the plants concerned, each species' sexual practice gets the job done.

Like any other manual, ours requires some ground rules and terminology.

The Scope of the Plants and Foods Considered in This Book

The focus is on the plants and their products that end up in our mouth as part of a meal, snack, or beverage. Some plant parts are picked directly off the plant and can be eaten immediately with little or no processing: fruits and their juice, nuts, seeds, vegetables, spices, and culinary herbs are examples. In contrast, the grains and dried legumes that provide the bulk of humanity's calories and the bulk of our macronutrients—carbohydrates, lipids, and proteins—almost always require some processing. Examples include rice and quinoa boiled into edible grain, wheat and rye ground and baked into bread, lentils soaked and boiled into dal, ground corn that creates tortillas and polenta, and buckwheat that ends up as kasha. The same is true for the underground plant parts that account for most of the rest of humanity's calories, such as potato, cassava, sweet potato (a species of morning glory; not at all a potato, which is a close relative of tomato), taro, and so on. The materials that create our stimulating beverages—coffee, *guaraná*, tea, cacao, and yerba maté—must undergo roasting, fermentation, or both. Fermentation is also a prerequisite for alcoholic beverages: fruit juice into wine, germinated and dried ("malted") grain into beer, agave into *pulque*. The additional step of alcohol-enhancing distillation creates products with a greater kick, for example, converting wine into brandy. Plant-derived products in our packaged processed foods often undergo so much manipulation that they bear no resemblance to the plants from which they came.

Check out the ingredients list on a bag of chips, a can of soda, or the wrapper from a chocolate bar, and see if you find any of the following plant products: high fructose corn syrup, corn oil, corn starch, sugar (comes from sugarcane or beets), soy lecithin, hydrogenated cottonseed oil, hydrogenated palm kernel oil, and cocoa butter. Those items are either relatively pure chemicals or simple chemical mixtures that are the products of processes that purify and/or alter plant materials into one or a handful of chemical compounds.

More challenging, see if you can find an ingredients list that doesn't include those chemicals. Keep in mind that humans are not the only organisms that process plant materials into foods. Think honey.

Every plant mentioned so far—including the tomato—belongs to a very large evolutionarily cohesive group called the flowering plants, which are characterized by flowers and seed-bearing fruits. The current scientific definition of a plant is much broader than that, including plants that disperse by spores instead of seeds or fruits (e.g., ferns), and plants that use cones instead of flowers to reproduce by seeds (e.g., pines). Sounds overwhelming? It's been worse. In the olden days (around when I was born), botanists used to consider "plants" to include all kinds of non-plant life-forms now known to be as evolutionarily distant (or even more distant) from plants as people are: bacteria, yeasts, algae, mushrooms, slime molds . . .

You might remember from high school biology that flowering plants are also known as angiosperms, referring to the fact that seeds (sperm) are borne in a fruit (*angio-*, meaning "closed container"). (It's weird that that their common name isn't "fruiting plants," but it's not.) With more than a quarter million known species and probably as many yet to be described, flowering plants are not only the largest group of plants; they comprise about 90% of the entire plant kingdom (Crepet and Niklas 2009). Angiosperms are so common and numerous that they are by far the most likely type of terrestrial plant you will bump into—and put into your mouth. More than 99% of the plant species that humans eat are angiosperms; flowering plants represent more than 99% of acreage of the food plants cultivated and harvested around the world, and they account for more than 99% of the calories that humans consume (directly or indirectly from the food chain).

Non-angiosperms have no fruits or flowers. The few significant edibles from such plants include pine nuts as well as fiddleheads from selected species of fern. All plants, flowering plants or not, have a life cycle that involves an alternation of generations. The complexities due to the alternation of generations are often important for non-angiosperms. Angiosperms have life cycles that are close, but not quite identical, to our own. Because our food is nearly exclusively angiosperm-based, for practical purposes let's dispatch with reference to alternation of generations at the moment. For example, I'll refer to flowers that bear only pollen as "male" because that is their function, rather than using the botanically correct (but clunky) phrase "male-gametophyte bearing." Lots of plant reproductive biologists do the same.

In conclusion, with regards to the basics of gamete production and fertilization, out of all plants, the sex life of angiosperms comes closest to that of animals but is far more exciting, as we are about to see . . .

Floral Terminology

Sex experts will tell you that to best understand how sex works, you need to know the parts and the positions. That's true for plants as well. The education involves some inevitable terminology. Fortunately, for the flowering plants, the four sets of floral parts are straightforward, if not intuitive. They are arranged coaxially as rings just like the circles in a bull's-eye. Better yet, think of a dinner plate painted with concentric bands.

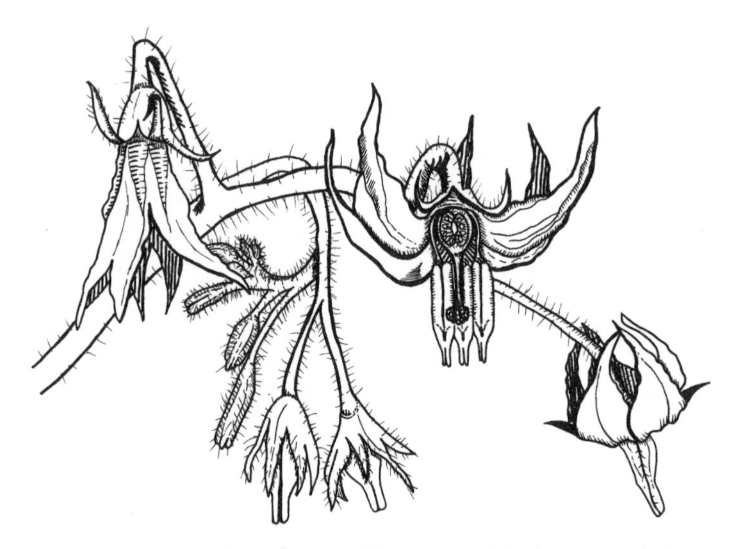

FIGURE 2.2 The tomato flower. Flowers at different stages of development attached to hairy stems. In most cases, the sepals, petals, and stamens are obvious. One mature flower is sliced in half from the side to expose part of the androecium and the inside of the gynoecium; all atop the perianth.

To learn the fundamental parts of the flower, take a look at the various tomato flowers in figure 2.2. Let's work from the stem end of the flower to the parts most distant from the stem. The stem end is also the outside edge of the dinner plate and the distal end is its center. The outermost/stem-end group is a whorl of sepals. In most cases, sepals are green and leafy. The green leafiness is a clue to the fact that all flower parts are modified leaves, molded by evolution for a reproductive function. In some cases, sepals are so showy that they might get confused with the petals. Finally, some plant species have dispensed with sepals altogether.

In the case of the tomato and members of its family, the sepals are always present, green, and leafy. In fact, it's a special feature of the tomato family that sepals persist after the associated fruit grows and matures, often sticking tightly to the fruit when it is harvested. In most other families, the sepals are long gone by the time the fruit matures. The next time you see an eggplant, a close relative of the tomato, take a look at the five green leaf-like things tightly attached to the stem end of the fruit. Those are the persistent sepals; collectively, the calyx of the flower that created the fruit. Indeed, if you find any fruit with five green leafy sepals tightly attached, you can be pretty confident that you've got another species in the tomato family. Chili peppers, anyone?

Inside the calyx is the next concentric set of parts, petals. The entire group of petals is the corolla. Like the sepals, the petals may be greenish and leafy. But more frequently, they range from pretty to dazzling; the corolla is the part of the flower that most people recognize as the flower. For some flowering plant species, the petals are absent (shocking!). Our tomato plant typically creates five showy yellow petals. In contrast to tomato family sepals, petals usually fall off the plant long before the fruit is ready for harvest.

The petals of many species are edible and can be tasty—though sometimes not. My own experimental sampling has taught me that the petals of lemon trees are pleasant and sweet, but the petals of the closely related mandarins are bitter. Foraging for petals in the wild is dangerous; eating the petals of certain species will make you sick or worse (e.g., larkspur). Also, it's wise (and obvious) to avoid eating the parts of wild plants that were possibly treated with pesticides or fertilized with raw manure (Newman and O'Connor 2009).

What about tomato flowers? The flowers of many species in the tomato family are known to be poisonous; I would assume tomato flowers to be the same. And that's all I wrote in my original manuscript.

When my manuscript returned with line edits, it included a comment from my editor, "But are they? What's the scientific verdict?" A smile of frustration blossomed on my lips. I had already spent hours entering the appropriate keywords (tomato, blossom, flower, toxic, edible, palatable, poisonous) first into Web of Science and Google Scholar to check the scientific literature. When I came up empty-handed, I repeated the search on Google. All I found there were websites that had the same expectation that I did: don't eat tomato flowers because the flowers of tomato relatives are poisonous. But no scientific sources were cited.

I asked a colleague, Professor Amy Litt, an expert on evolution and development in the tomato family. She reminded me that she is no expert on tomato but guessed that maybe the blossoms are edible. She also recommended that I ask the world's go-to person in botany on all things tomato and tomato family. That scholar wrote back that she couldn't answer the question and was not aware of whether relevant research had ever been done.

Now my editor had reopened the box of tomato worms. Back to more hours trying new combinations of words online. I found a recipe for "tomato flowers" with instructions on how to cut the fruit into a blossom-like form. Back to Amy Litt, who decided to use the world of social media. Her Facebook query drew a lot of attention and two useful responses: One of her ex-students was aware that a tomato relative—the sometimes edible, sometimes poisonous European black nightshade—had flowers that did not bear any of the typical alkaloid toxins found in some other parts of the plant. "Probably edible" was the conclusion. The other was an anecdotal response from a friend whose boyfriend had consumed tomato flowers in his youth.

While waiting for Amy's social media to work its magic, I had asked around some more. I randomly queried about a half dozen other plant scientists on campus. "Probably poisonous," they all agreed. I sought out the advice of two highly trained gourmet chefs who happen to be my friends. They gave almost identical responses: "Norm, you should already know that tomato is in a plant family famous for its poisonous species." I even asked an owner of a company that grows, packs, and distributes tomatoes. Her response: "Why would anyone want to eat tomato flowers when they can eat the fruit?"

I'm no longer frustrated because I can use this adventure to illustrate the reality of science. Scientists don't know everything. More importantly, they haven't even gotten around to conducting the research necessary to answer all of the seemingly obvious questions.

"What's the scientific verdict?" The jury has not yet returned; indeed, it hasn't even heard the evidence!

Together the calyx and corolla form the perianth. In the tomato family, it's easy to tell petals from sepals. But for other plants, like lilies, sepals and petals look pretty much the same, usually forming a six-part perianth of three showy petals and three showy sepals. When you eat daylily buds ("golden needles") in Chinese soup, most of what you eat is perianth. Even though the perianth is botanically considered as a set of reproductive organs, and as pretty as the perianth may often be, the parts are sexual advertisement at best—the wardrobe and makeup. The two inner whorls are the sexual business end of the flower.

Like most plants, the tomato involves flowers that have both male and female parts. Botanists, as you will soon learn, have a penchant (bordering on fanatic) for creating terminology. I'll be your filter, giving you just enough for your use and, occasionally, amusement. (Be grateful for all of the filtering that I've already done for you!) Flowers with both male and female parts are termed bisexual or hermaphroditic. The third botanical name for such flowers is rather sweet: perfect.

Within the corolla are the male parts, a whorl of stamens. Typically, stamens are assembled with two structures, the pollen-releasing part, the anther, sitting atop a stalk-like filament. Filaments can range in size from longish to inconspicuous. Collectively, stamens—structures of the flower associated with male function—are called the androecium (*andr-* plus *oecium* from the Greek *oikion* for, more or less, "man's house"). Depending on the species, a flower may have only a single stamen or up to a hundred or more. Likewise, and also varying among species or individuals, some flowers are male-sterile, lacking stamens altogether. Some species have both fertile stamens and sterile ones that develop into attractive structures called staminodes. Wild roses have five petals and loads of stamens. But if you get roses from your sweetie, you'll count a lot more than five petals because rose breeders have genetically "improved" the rose by selecting for types in which some of the stamens have evolved into convincingly petal-like staminodes. (Note: no genetic engineering involved.)

Functioning stamens are the rule for tomato flowers. The five stamens are shaped and colored like bananas. They are fused with each other to form a cylinder around the female organ. In the tomato, the filaments are so short as to be hardly noticeable.

Within the encircling androecium is the bull's-eye of the

flower. The gynoecium (the "woman's house") is also known as the pistil. (Botany joke: Why are field botanists on collecting trips safe from bandits? Because they carry pistils [Larry Venable, personal communication].) A gynoecium conducts the female business of the flower. Like the androecium, it is a set of one or more parts; in this case, carpels. This organ is often shaped like a vase with a longish neck and a flared top. The portion of the carpel that receives the pollen, the stigma, is typically the surface of that flared tip. In Food World, the most famous stigmas are the three long threads that emerge from the pistil of the flower of *Crocus sativus*. Hand-harvested and dried, they become the spice saffron, essential for *paella Valenciana*. Quality saffron is reputed to be the most valuable food by weight. I checked its price against silver today (December 1, 2017): an ounce of high-quality saffron at retail will buy you about sixteen ounces of silver or two-tenths of an ounce of gold.

A pollen grain that ends up on a stigma germinates and then grows a tube that makes its way down through an extended neck-like portion of the carpel called the style. (Harvesting both stigmas and styles of *C. sativa* is much easier than the stigmas alone, but yields lower-quality saffron.) The tube eventually arrives within the expanded bottom of the gynoecium, the ovary. There, one or more egg cells are waiting to be fertilized by sperm cells delivered by the pollen tube. After fertilization occurs and seeds begin to develop, the fertilized carpel or carpels develop to become the container that holds the plant's eggs, the fruit. Many fruits, from avocados to cherries to peapods, develop from a single carpel. But most fruits are the result of two or more fertilized carpels fused together.

The tomato is an example of the latter. The tomato's

TABLE 2.1. *The Four Types of Flower Parts*

Individual part (starting at the base of the flower)	Number of parts in tomato flower	Collective whorl	Functional sex part?
sepal	5	calyx	no, sterile
petal	5	corolla	no, sterile
stamen	5	androecium	yes, male
carpel	2	gynoecium (aka pistil)	yes, female

vase-shaped green gynoecium is a single organ composed of two fused carpels encircled by the stamen cylinder. If you cut a mature tomato crosswise, the various invaginations of the fruit make those carpels a bit hard to see. Take a look at the tomato cross-section on the book's frontispiece. You will recognize the thick wall that bisects the fruit, segregating the two carpels. In citrus fruits, the carpels are very apparent, what we commonly call segments. Try cutting a cross-section of an orange and see if you can count the ten or so carpels. Flowering plants with multiple unfused carpels developing from single flowers are rare. Nonetheless, we appreciate some of these as food sources. For example, a single raspberry develops from a single flower but is a cluster of individual fruits, each developing from a single carpel.

Even though the United States Supreme Court determined tomatoes to be vegetables (Jones 2008), just about everybody knows they are actually fruits. Conversely, some commonly called "fruits" are simply not. More than 90% of what we call a "strawberry" is a product of something that was never part of the flower. The red, sweet, juicy thing that develops after fertilization is the tip of the stem adjacent to the calyx. The botanically defined fruits of the strawberry plant are those little brown things that we typically call seeds. Yes, there's a seed inside of each of those tiny hard fruits, each one developing from an individual carpel of the strawberry flower.

Sex by the Numbers

The father of modern botany, Carolus Linnaeus, spent more time thinking about sex than any other botanist of the seventeenth century. He pondered the numbers of flower sex parts ("She loves me, she loves me not . . ."?). The Swede noticed that the number of sepals, petals, stamens, and carpels is remarkably similar for plant species that share other characteristics. He recognized this pattern as the key to organizing the diverse world of plant species.

The typical tomato flower has five sepals, five petals, five stamens, and two carpels. This 5-5-5-2 floral formula holds for the flowers of lots of other tomato-like food plants: eggplant, tomatillo, tamarillo, goji berry, potato, and so on. The floral formula of the tomato and the eggplant is shared by petunia, nightshade, tobacco, jimson weed, as well as thousands of other species; Linnaeus placed them all in the same plant family. Biology 101 teaches us that Linnaeus invented the two-name system of naming species, both animal and plant. Tomato is *Solanum lycopersicum*: genus *Solanum*, species *lycopersicum*. But his observations regarding the commonality of the numbers of plant sex parts are far more significant. They kicked off his earth-shaking organization of species—first plants and eventually animals—into hierarchical groups. Groups of species are organized into genera (singular, genus). Groups of genera are organized into families.

Initially, Linnaeus used the floral formula as a primary tool for assigning plants into families. Tomato's family is Solanaceae. Using the floral formula 5-5-5-2 and one or two other characteristics, such as persistence of the sepals that are fused to each other, it is pretty hard to fail to identify a plant in the Solanaceae family. The thousands of solanaceous species have commonalities beyond flower and fruit characteristics. In particular, all contain various amounts and various types of a special class of chemical compounds, alkaloids. These include the nicotine in tobacco, the capsaicin that gives chilies their heat, and the chemicals that account for both the toxic and medicinal properties of various nightshade species. While alkaloids are a characteristic that is shared by members of the Solanaceae, some species scattered over other unrelated families create famous alkaloids such as caffeine and theobromine.

Sexual Diversity

Now that we've got the sex parts down, let's move on to what the plants do with those parts: the dizzyingly diverse world of plant sexuality. As mentioned above, flowering plant species, including those that we eat, most commonly have perfect (bisexually functioning) flowers and can self-fertilize. But even self-fertilization isn't as simple as it sounds. Many self-fertile plants can accomplish self-fertilization in two different ways.

Our tomato can illustrate. A flower can pollinate and fertilize itself. Also, different flowers on the same plant can successfully mate with one another. In a self-fertile plant, both ways yield the same result, offspring mothered and fathered by the same individual. In the case of the tomato, it is necessary for an outside agent (a pollinator) to get the job done.

In the wild, the tomato's ancestors were (and are) self-sterile, requiring an insect, usually some species of bee, to move pollen from plant to plant. The process of getting the pollen out of the anthers, famously known as buzz pollination, depends on the insect physically vibrating the flower. The pollen grains emerge from the anthers like toothpaste from a tube. For a good set of nicely shaped fruit, tomato flowers still need to be buzz-pollinated. When natural bees are in sufficient abundance in the field, pollination is no problem. But what about winter-grown hothouse tomatoes? Bumblebees can be employed for this process, or (and I do not make this stuff up) farmers can use electric vibrators (Jones 2008). I might add that such vibrators are specifically designed with the tomato in mind.

The fraction of seeds produced from self-fertilization (aka selfing) versus mating with another individual (outcrossing) varies tremendously among plant species. Most angiosperms are self-fertile, but many of them go *way* out of their way to outcross (Richards 1997). Even though most plant species have the opportunity to self-fertilize, the vast majority of the self-fertile have mechanisms that encourage interplant mating. Interplant romance rocks. Species with high levels of selfing are in the minority. One hundred percent selfing is almost unknown in the plant and animal kingdoms. Charles Darwin (1885) pithily observed that nature "abhors perpetual self-fertilization." Darwin (1876b) wrote an entire book on this phenomenon: *The Effects of Cross and Self Fertilisation in the Vegetable Kingdom.* He was as interested in plant sex as he was by the evolution of the species that create our food (see *The Variation of Animals and Plants under Domestication* [Darwin 1868]).

Our most important food plants buck the common trend of intermediate and high levels of outcrossing. Of the world's top ten crop plants, all but one (corn, aka maize) produce more than half of their seeds via self-fertilization, and more than half of those produce more than 95% of their seeds from selfing: wheat, rice, soybeans, barley, pearl millet, and beans (Andersson and de Vicente 2010). The same is true for lots of other food plants, including our beloved tomato, as well as cowpeas, chickpeas, and peanuts. Many other food plants are considered "mixed mating" with intermediate (10–90%) outcrossing rates. Avocado, canola, and grain sorghum fall into this category of food plants that produce a mix of outcrossed and selfed seeds.

Although 100% selfing is extraordinarily rare in plants, 100% outcrossing is not. In accord with Darwin's observation, the second most common type of plant breeding system—self-incompatibility—works to prevent mating with oneself. Such plants *must* outcross to produce seeds. Typically, self-incompatible plants have perfect flowers, but if a pollen grain produced by such a flower finds its way onto a stigma of the same flower that produced it, self-fertilization is physiologically blocked, and it is blocked for all other flowers on the same plant. For certain self-incompatible crops, like radish and rye, the pollen grain cannot even germinate on an incompatible stigma. For others, a self-produced pollen grain will germinate on the stigma and grow into a style but will cease growth long before the pollen tube reaches the egg (e.g., certain citrus [Kahn and DeMason 1986]). Self-incompatibility still leaves a lot of opportunities for mating.

The most common form permits mating with just about everyone else in the local population, except for oneself and some very closely related individuals (Richards 1997).

Species comprised of individuals incapable of mating with themselves are considered obligate (100%) outcrossers. The previously mentioned rye and radish, as well as almond and cabbage, are obligate outcrossing crops because of self-incompatibility; there are many more. Nonetheless, self-incompatible food plants are relatively uncommon compared to their representation in the natural world of wild species. Many of our self-fertile food plants were domesticated from self-incompatible ancestors but picked up genes for self-compatibility along the way. Sexy, but self-fertile, the tomato has a self-incompatible wild progenitor; the same is true for cultivated rice and its ancestor (Rick 1988).

Self-incompatibility is not the only mechanism for obligate outcrossing. Some species have individuals that express only one sexual reproductive function or the other. Does this arrangement seem familiar? It should be; it's ours. (It's about time!) Separate sexes, so common in the birds and the bees—as well as mammals, reptiles, and fishes—is known as dioecy (*di-* = two; *oecy* = houses); the derived adjective is dioecious. In a dioecious plant species, all plants produce only "imperfect" unisexual flowers. Some plants solely express female function; all the other plants create flowers exclusively expressing male function (see table 2.2). In other words, populations are composed of functionally male individuals and functionally female individuals. Genuine dioecious flowering plant species are rare, roughly 5% of the total (Renner 2014). This breeding system characterizes a diverse set of food plants. Date palms come as male and female trees. Likewise, kiwifruit vines, mulberry trees, and

TABLE 2.2. *The Sexuality of Flowers Comes in Four "Flavors"*

Functional sex parts present	Type of flower
Male and female	Perfect, aka bisexual, aka hermaphroditic
Male only	Male, aka staminate, aka female-sterile
Female only	Female, aka pistillate, aka male-sterile
No sexually functional parts	Sterile

asparagus are dioecious species. The same is true for hops and its extremely close relative cannabis. The store-bought strawberry is produced by plants with perfect flowers, but its wild ancestors (also edible) are dioecious.

Now for wilder territory. Corn, by any measure, ranks as one of the world's top three crops. A common feature of New World culture, the corn plant is easy to visualize. A single thick stalk topped with a tassel has long strap-like leaves with a few large ears hanging off the sides. The imperfect flowers that hang from the tassel on top are male; the ones with the "silks" (long stigmas) coming out of the leaves that enclose the ears are female.

An individual that produces unisexual flowers of both types is said to be monoecious. Monoecy ("one house") is much more common in plants than dioecy. Like dioecy, under monoecy, individuals of the species never produce perfect flowers. But, unlike dioecy, all plants are of a single type, with both male and female flowers found on the same plant. Monoecious plants are bisexual but contrast with the much more common state of bisexual hermaphroditism in angiosperms, in which all plants are of a single type but male and female functions coexist in perfect flowers.

Corn is the most important monoecious food plant. A century ago, those who studied corn assumed it was gravity-pollinated; that is, pollen falling from the tassel to the silk caused self-fertilization, a misconception that persists into modern times (e.g., Pollan 2006). But once molecular genetic analysis of corn seed was available for measuring outcrossing, geneticists discovered that corn, like nature, loathes high levels of selfing. Despite its self-compatibility, corn outcrosses at levels in excess of 95% (Bijlsma, Allard, and Kahler 1986).

Edible plants that enjoy monoecy are abundant in the cucumber family. The next time you sink your teeth into a cheese-stuffed, deep-fried squash blossom, reflect on the fact that the blossom can be either male or female (most likely to be female) but cannot be both. The first flowers on squash vines are male; the plant must reach a critical size before it starts producing female flowers.

You will note that it is a lot harder to self-fertilize if your male and female flowers are separated on different parts of the plant. Botanists (with their penchant for creating terms) have named the spatial separation of sex expression "herkogamy." ("Herk"?) Just as plants can divide their sexual functions in space, they can also separate them over time.

Dichogamy (pronounced dye-COG-a-mee, not DICK-o-gam-y) is the name for the separation of plant sex expression over time. Male-first dichogamy, as practiced by the squashes and their relatives, is protandry (essentially "male first"). Female-first dichogamy (protogyny) is much rarer in the plant world. Thus, the monoecious species of the cucumber family are not only herkogamous (male flowers mostly on the older part of the vine), but they enjoy whole-plant dichogamy as well (female flowers after male flowers). Will you ever look at a pickle in the same way again?

For the moneocious cucumber family species, dichogamy

occurs as a whole-plant phenomenon. For certain species in other families with perfect flowers, the flowers themselves can be dichogamous, expressing one sexual function first and then the other, as we will see later; chapter 4 is entirely devoted to a food plant that practices that kind of protogyny.

And yes, perfect flowers can be herkogamous as well. As you cut into your stack of buckwheat pancakes, thank within-flower herkogamy for your meal. The buckwheat species is composed of two different kinds of herkogamous plants that differ with regard to how their pollen-bearing parts and pollen-receiving parts are spatially separated in the flower. Both kinds of plants have perfect flowers. One type has flowers with a long-styled gynoecium that dwarfs the shorter stamens; for flowers of the other, the short-styled gynoecium is shadowed by the longer stamens.

Such "heterostylous" species were a favorite of Darwin (1876a), who described their properties in his book *The Different Forms of Flowers on Plants of the Same Species*. You already know the other two types of plant species that have different forms of flowers (dioecious species and monoecious species, right?). Monoecy and dioecy are child's play compared to heterostyly's lock-and-key-like sexual complexity. Here's how heterostyly works. In most heterostylous species, including buckwheat, long-style plants have a type of self-incompatibility that prevents self-fertilization and prevents successful fertilization with any other long-style plants; likewise, short-style plants cannot self-fertilize, nor can they successfully mate with others of the same morph. Long-style plants successfully mother and sire the children of short-styled plants and vice versa. It is worth making it clear: heterostylous species are composed of plants that can be assigned to two morphs, *but not male and female plants*, because both kinds of plants have functional male and female parts within their flowers.

Heterostyly is a rare plant-mating system, concentrated in a few plant families and known from several hundred species. I am aware of only one other food plant that has heterostyly, oca, the second most important staple crop of the Incas (after potato). Still an important crop of high-altitude South America, oca is now a cultivated commodity in Mexico and New Zealand. The plants of this tuber crop come in three types—long style, mid-length style, and short style—each incompatible with themselves but compatible with the other two. The oca's three-morph heterostyly is named tristyly to distinguish it from the more common two-type system, distyly, typified in buckwheat.

Now you've learned your basics of plant sexuality for the vast majority of our food plants. But you've only advanced to intermediate level. Heterostyly gives a taste of advanced-level plant sexual systems. My book, your attention, and our mutual sanities would be strained to thoroughly detail every complex method of plant mating.

With the tools you have accumulated in the last few pages, you now have the expertise to dig deeper into the dozen or more other variants of plant sexuality. Here's a partial menu: gynodioecy, heterodichogamy, trioecy, polygamodioecy, cleistogamy, andromonoecy, permanent odd polyploidy, and enantiostyly. Some of these are explained in table 2.3. To do a complete detailed inventory would fill a book on its own. Readers who cannot get enough will be thrilled to learn that such a book is available: *Plant Breeding Systems*, by A. J. Richards (1997).

TABLE 2.3. *Representative Types of Flowering Plant Sexuality*

Distribution of flower types over individuals	Name	Food plant example	Comments
All individuals bear only perfect (bisexual) flowers.	hermaphroditism	tomato	Most common for food plants
All individuals can bear both male and female unisexual flowers, but do not bear perfect flowers.	monoecy	corn	Similar to earthworms, many shellfish, and some other animal species
All individuals bear both perfect and female flowers.	gynomonoecy	sunflower	
All individuals bear both perfect and male flowers.	andromonoecy	carrot	
Two kinds of individuals: some bear only male flowers; others bear only female flowers.	dioecy	date palm	Similar to humans, birds, mammals, most insects, and many other animal species
Two kinds of individuals: some bear only perfect flowers; others bear only female flowers. Or some bear both male and female unisexual flowers; others bear only female flowers.	gynodioecy	thyme	
Two kinds of individuals: Some bear only perfect flowers; others bear only male flowers. Or some bear both male and female unisexual flowers; others bear only male flowers.	androdioecy	none	Exceedingly rare in both plants and animals

Pollination Mechanisms

We've nibbled around the edges of how plants get pollen moved from anther to recipient stigma. Let's take a closer look starting with the simplest case: plants that are considered highly self-fertilizing, that is, typically outcrossing at rates of 5% or less. Highly self-pollinated plants have their stamens and stigmas arranged sufficiently close for a possible direct handoff of pollen to the stigma within a flower, allowing pollination either in the bud or soon after flowering.

Certain other self-fertile, but otherwise outcrossing, species have the opportunity for delayed selfing. The male and female parts are initially at a distance from each other but slowly enlarge or change shape after the flower opens such that they eventually come into contact hours or days after the flower opens. If pollination has not occurred, the flower may self-pollinate. Plant reproductive ecologists consider this delayed selfing as "fertilization insurance," just in case the plant doesn't receive pollen from another mate. As in the case of the tomato, yet other self-fertile plants need help to self, with the aid of the wind or an animal that disturbs the flower.

So far, we have only considered the kind of self-fertilization that occurs within an individual flower, autogamy. Self-compatible plants have a second option for mating with themselves. If you watch your tomato plant, you might see a bee working one flower after another before it moves on to another plant. Because the tomato is self-compatible, it's likely that the bee moving pollen among flowers within the same plant ends up successfully pollinating those flowers. There's a term for between-flower self-pollination: geitonogamy. Depending on the species, either the wind or an animal can do the job.

Wind and animals can be the agents of outcrossing as well. A tiny fraction of flowering plants depend on water to carry pollen from flower to flower. A few other proposed mechanisms of pollen transfer remain speculative. For example, the hypothesis that rain acts to fertilize the plant that produces black pepper is controversial (Cox 1988). The effects of other pollen vectors are almost nil for the plants that produce our food.

"Grasses have flowers," the biggest shock of high school biology. *That's just wrong!* The same response came from my biomedical scientist father-in-law when his botanist daughter told him the secret. The truth is that grasses do have flowers, but they are tiny.

Grasses, the family Poaceae, are the menu item responsible for most of humanity's calories, in the form of cereal grains. Four of the world's top five crops are grasses: corn, wheat, rice, and barley. A few grass family food species are grown exclusively for parts other than grain, such as the stalks of sugarcane.

Animal pollination is almost unknown in the grasses. Some cereal species create seeds mostly by selfing: wheat, oats, barley. Some, like sorghum, employ a mixture of wind and self-pollination (Ellstrand and Foster 1983). Self-incompatible rye is an example of a 100% wind-pollinated cereal crop. Wind pollination, anemophily ("love of the wind"), requires massive production of tiny, buoyant pollen grains. After all, the wind doesn't particularly care where it places those grains. It's not unusual for a single plant to create millions of pollen grains in one season. A single walnut tree releases considerably more a *billion* grains into the air (Molina et al. 1996). Obviously, the vast majority die, lost and alone. A few get lucky, finding their way to a receptive stigma. If you ever have the opportunity to encounter a corn plant when its tassel is releasing pollen—just after the dew returns to its source—give it a hand. Whap the stalk with your finger and watch the pollen cloud take off. Wish those guys good luck.

Outside of the grass family, some additional food plants depend on wind pollination to create at least some of their seeds and fruits. Grapes are one example, being both self-pollinated and wind-pollinated. Wind pollination is the rule for edibles as diverse as beets, spinach, hops, and hazelnuts. Mulberry trees engage in wind pollination with gusto; they explosively release their pollen—at more than half the speed of sound (Taylor et al. 2006).

Animal pollination comprises diverse actors from the birds and bees to thousands of other insect species. Likewise, the mammals that fly, bats, and a handful of those that don't—from marsupial honey possums of Australia to lemurs in Madagascar—assist in pollination. Plants that use animals as agents of pollination usually provide both advertisement and reward. Advertisement can be visual (a big and bright floral display) and/or olfactory (scent). Rose and violet flavors have been recently popular in haute couture ice creams and candies. The flavors are extracts of perfumes that these plants employ to entice pollinators. If you don't want to get so exotic, try orange blossom honey versus buckwheat versus clover to sample some of the spectrum of scents. As processed by honeybees, nectar converted to honey often retains the scent of the flower. If you taste them side by side, orange blossom honey and buckwheat honey are so radically different that they seem like totally different foodstuffs.

Reward often comes in the form of sugar-rich nectar

and/or protein-rich pollen. Less frequently, flowers offer up oils as a reward to pollinators, but those rewards are not prominent in the plants we use for food. For food plants, animal pollen vectors are overwhelmingly insects. Honeybees are the best known pollinators of flowering plants and assure seed set for foods from almond to alfalfa. But for some crops, bees are either inferior to other insects or even totally ineffective. Beetles are the pollen vectors of choice for most of the custard apple family that produces an array of important tropical and subtropical fruits from cherimoya to sweetsop. Some cultivars of figs require pollination by tiny specialized fig wasps. Mammalian pollination is critical for one commercially important food plant: the date palm; without humans delivering pollen from male date palms to female trees, fruit set is negligible.

Some species can utilize all three common pollination modes to get the job done: self-fertilization, wind, and animals. Sexually versatile yellow-flowered rapeseed plants (whose products include canola oil) are particularly promiscuous. They can self-pollinate, receive pollen via the wind, and accept pollen from bees, flies, butterflies, moths, and other insects. For rapeseed, all three types of pollination can lead to successful fertilization.

Asexuality Rears Its Ugly Head

Floral architecture, nectar, dichogamy, self-incompatibility. Clearly, flowering plants often go to great lengths to make sure that they engage in romance. Well, it makes sense, doesn't it? After all, no sex, no reproduction, right? No. It's time to share a secret and a great mystery. The secret: While the great majority of flowering plant species have the option of sexual reproduction, a great many, probably the majority of angiosperms, are able to reproduce *without* sex, via asexual reproduction, also called apomixis (*apo* = without; *mixis* = mingling). The mystery: Why sex?

Make no mistake. Asexuality is *not* the same thing as self-fertilization. Selfing requires the creation of gametes—egg and sperm—and their fusion for fertilization. Each gamete is a genetic subsample of the parent. In the case of self-fertilization, the genotype of the resulting child may be similar to the parent but is far from identical. Self-fertilization is sex.

Then what is sex? Sex is one of the processes that create genetic novelty. Most readers will recognize that, in the case of humans, sex is intimately linked to the process of reproduction:

"Wow, she has her mother's nose and eyes!"

"Yes, but she has her father's ears."

"Do you think she'll have her mother's skill with trigonometry?"

"Hard to say, but let's hope she will be able to barbecue like her dad."

The basics of human heredity are straightforward. Children, the product of sex, share some characteristics with each parent but are never identical to a parent. Some hereditary information from each parent gets mixed and transmitted to the next generation. Geneticists call this mixing recombination, or simply sex. The scientific definition of sex (depending on *which* science you choose—in this case, the various genetics-oriented sciences, including evolution, population genetics, plant breeding, etc.) is an organism-organism interaction resulting in some mixing of genetic material, producing one or more organisms that are genetically different from the original sources. (Note: Reproduction is not part of the definition of sex. But let's revisit this surprising fact in chapter 6.) In the case of human sex, the organism-organism interaction is the fusing of two single-cell gametes—one egg and one sperm—creating a zygote with a novel genotype. It's pretty much the same for lions and tigers and bears and food plants. For self-fertilization, whether practiced by a tomato plant or a banana slug, the egg and sperm are created by the same individual; the organism interacts with itself.

Gamete creation pulls a different sample of genes out of the parental bag. Given two copies of tens of thousands of genes per individual, statistically speaking, the gametes that fuse are different from one another and different from the others produced by that individual. To paraphrase Fred, aka "Mister," Rogers, every child, whether from selfing or outcrossing, is unique.

No so for the asexual child. As true to the original as a photocopy, the asexual child matches its mom. No dad needed. The resulting progeny is a clone, a genetic copy of the parent. In plants, apomixis occurs by one of two methods.

Some species are able to reproduce vegetatively, without depending on flowers, fruits, or seeds. Most of these engage in sex and reproduce by seed as well. Strawberry plants can spread and create new plants by aboveground stems called runners. If the runners are cut or decay, the resulting units are separate individuals. And indeed this is how we propagate them. The tuber we call a potato is a belowground stem that can create new tubers and new plants. It is well known that a chunk off a potato that includes at least one "eye" can be planted to grow a new plant. A prickly pear cactus pad—if not peeled, cooked, and eaten as a nopal—can be treated the same way. Clonal propagation is the standard way that many of the world's most important perennial crops are multiplied: in addition to strawberry, prickly pear, and potato, such crops include sugarcane, banana, cassava, plantain, and sweet potato.

Other species create apomictic offspring by seed. Instead of first creating an egg cell that represents half of the mother plant's genetic material and waits for a sperm to start on its way to a seed, the asexual seed develops as a genetic copy of the mother with a full complement of her genetic material. The lowly dandelion that sometimes makes its way from the lawn into salads and wine is the scientist's poster child for apomictic seed. The seed created by the dandelions in your yard are copies of their moms.

The situation isn't so simple for citrus fruit species. Some varieties, like the Valencia orange, the Marsh grapefruit, and the Dancy mandarin follow the dandelion model, producing all or nearly all seeds asexually. In contrast, the seeds of cultivated varieties of another citrus species, the pummelo, are solely the result of outcrossing. Still others, like the Mexican lime and the Eureka lemon, produce a mix of sexual (both selfed and outcrossed) and asexual progeny. Citrus trees aren't capable of vegetatively propagating themselves, but humans can do the job for them. By grafting buds onto rootstocks, Californians have multiplied and re-multiplied millions of trees that produce the seedless Washington navel orange—all descended from three budded trees brought to Riverside, California, in 1873.

Most plant species that are capable of reproduction without sex include sexual reproduction as an option, most often as the preferred option. The developmental pathways to asexuality can be as easy as simply falling apart. Not so for sexuality, especially outcrossing. But even the highly selfing tomato builds bright yellow flowers and produces much more pollen than necessary for self-pollination. Given that plants have to go to great lengths to engage in sexual reproduction, we return to the mystery of "Why sex?"—or more precisely: "Why does the hassle that is sex persist?" To answer that question, we need to take a long and hard look at life without sex in a crop that produces a long and hard fruit, in chapter 3.

Sweet and Savory Valentine Pudding

Despite the fact that Thomas Jefferson grew and enjoyed tomatoes at his table, they were regarded by most early Americans with suspicion. It is said that the courageous Robert Gibbon Johnson changed the course of history in 1820 when he announced that he would eat tomatoes on the steps of the Salem, New Jersey, courthouse (Rick 1978). When the throngs anticipating a public suicide arrived, I suspect they were variously disappointed, confused, or thrilled to see no ill effects. In any case, we in the United States are better for his bravery. "From pizza to the Bloody Mary" (Rick 1978), Americans have embraced this Latin American immigrant as their own.

Tomato pudding is a sweet and savory holiday dish for all occasions. Like the tomato, my recipe for tomato pudding has a long and winding history. As far as I can tell, the recipe originated somewhere in the southeastern United States within a decade or so after Robert Gibbon Johnson's public experiment.

My mother-in-law, Judith Kahn, enjoyed this sweet tomatoey side dish at a hotel restaurant near the University of

Michigan Alumni Camp year after year in the 1960s. She finally asked for the recipe. She has used this pudding as a tasty counterbalance to holiday meats for Thanksgiving dinner. I've modified the following recipe to bring out the flavor of the tomato (that is, to attenuate the 1960s calorie-laded, dessert-like sweetness of the inherited instructions). When you prepare and enjoy this one, remember the triumph of the love apple over its undeserved reputation for being toxic.

10 ounces tomato puree
½ cup water
½ cup packed brown sugar
1 teaspoon salt
1 tablespoon molasses
2 cups dried cubes of good bread
½ cup melted butter

Serves four to six as a side dish.

Preheat oven to 400° F. Add the water, brown sugar, molasses, tomato puree, and salt to a saucepan and bring to a boil. While those are heating, place bread cubes in a layer in a one-quart ovenproof greased casserole dish and cover with the melted butter. Pour boiled mixture over the buttered cubes. Bake 45 minutes. Serve warm. Works well with roast turkey, ham, lamb, or tempeh. A doubled recipe yields a pleasant leftover.

3 *Banana*

A LIFE WITHOUT SEX

Yes! We have no bananas . . .
—lyrics by Frank Silver and Irving Cohn (1922)

The banana is arguably the world's sexiest fruit, a conclusion that, in this PG-13 volume, warrants no further discussion. The banana has earned other superlatives as well. Dan Koeppel (2008) called bananas "the world's most humble fruit." That's quite a difference from the Chiquita bananas' website (www.chiquita.com), which proclaims them to be "quite possibly the world's most perfect food." And why not? Not many foods come ready to eat in their own natural biodegradable container. Nutritious? You bet. A single banana provides much more than 10% of your daily potassium, fiber, vitamin C, and manganese, plus 20% of vitamin B6—with essentially no fat, cholesterol, or sodium.

The banana is the world's most important fresh fruit crop, a big deal if you are up against competition like citrus fruits, grapes, and apples. Furthermore, this statistic is for the banana alone and does not include its close relative, the starchy plantain, which must always be cooked to be consumed. Grapes have far more total acreage and production than bananas, but after accounting for those grapes allocated for processing into juice, wine, and raisins, a smallish fraction is left to be eaten out of hand as fresh fruit. Globally, apples come close to bananas in terms of acres planted, but again a substantial portion of the apple crop gets juiced, with the real thing, more often as concentrate, ending up in drinks that don't even allude to "apple" in their name. Try a packaged raspberry or even grape "drink" hinting that fruit is involved in the mix. Then check the ingredients list for a surprise.

Only a teeny portion of the banana crop gets processed for things like baby food or dried banana chips. Some bananas get cooked like their plantain cousins. Others find their way into beer in East Africa. But among fresh fruits, the banana is, indeed, the Top Banana. Although a lot of bananas remain at home for consumption in, for example, Brazil, India, and parts of Africa, vast quantities are grown for export from tropical countries for fresh consumption in economically developed countries. At the moment, the EU, the United States, and the Russian Federation are the top importers. Ecuador is the top exporter, followed by the Philippines, Guatemala, Costa Rica, and Colombia.

(A brief ag/food statistics digression and suggestion: Where do I get these rankings of crop production per acre and trade statistics? If you cannot get enough of facts such as how much a given country produces, processes, or imports agricultural products, by-products, and associated products like wheat, cheese, ghee, hops, mules, bananas, forest trees, money, people, molasses, yerba maté, or greenhouse gases, then your go-to website is FAOSTAT: www.fao.org/faostat/. The Food and Agriculture Organization of the United Nations [FAO] runs this user-friendly, accessible, half-billion-byte and growing database. It is utterly addictive for certain dispositions. *Now excuse me while I look up the second most important agricultural product of Djibouti, as ranked by weight. . . . Hmmm? . . . Camel meat. . . . Oh, how cool!*)

The banana was the first industrial fruit. The sociopolitical relationship between the banana-producing "banana republics" and banana-consuming nations has led to interesting and sometime infamous turns of history; to name a few, environmental degradation, depletion of biodiversity, exploitation of workers, unsafe pesticide use, social disruption, and political instability. The outcomes, for better and worse, are still being played out on the world stage. Several recent books have been written on how the banana changed the world (e.g., Frank 2005; Chapman 2007; Koeppel 2008; Frundt 2009).

Our interest is in how the world changed the banana. The mention of sex, that genetic variation machine, leads to yet another superlative. Of the important global crops, the banana is the most genetically uniform. A single cluster of nearly identical genotypes, the Cavendish subgroup, nearly monopolizes the world's banana groves and banana trade. In contrast to the riotous rainbow of genetic diversity that lends sustainability to natural plant and animal populations, the world's banana industry has the stability of an upside-down Egyptian pyramid balanced on its tip. That fact leads to the final superlative: the commercial banana is the world's most endangered major crop. The future of the intercontinentally traded banana was once, and is again, precarious. Given that their wild progenitors are as variable as most species, how has it come to pass that most of the banana plants growing in the world have become so uniform? And what does that uniformity mean for their future as the "world's most perfect food"?

Commercial industrial agriculture has stolen sex from the banana.

To understand the banana's future, we need to understand the origins of the supermarket bananas that we know today. Short of the use of a time machine, the detailed origins of a crop will always involve some speculation. Scholars originally supposed domestication of any given crop to be the singular result of some innovative Neolithic proto-farmer shoving a few seeds in the ground and striking it big. Cotton—comprised of four different species, each domesticated in four far-flung locations: Central America, the Middle East, South America, and India—was the weird exception (Wendel 1995). In the last twenty years, scientists—using the combined tools of new genomic and archaeological analysis—have come to recognize that entrepreneurial brilliance was more common among Neolithic people than previously guessed. Evidence for multiple independent domestication events has rapidly accumulated for many crops (Mayer and Purugganan 2013). Soon it may be shown to be the rule, rather than the exception.

The banana falls into the category of a good idea whose time had come to different folks in different places. The banana (not counting the look-alike plantain) was domesticated at least twice in the tropical region extending from Malaysia through Indonesia and New Guinea and into the Pacific (Simmonds 1995). The banana's wild forebears remain in the region to this day. The domesticated banana plant and its ancestor have a lot in common. Both wild and domesticated bananas produce nearly identical flowers. Grown side by side, one substantial difference separates them. The fruit of the wild banana is practically inedible.

Wild bananas and cultivated ones are big perennial herbs that are superficially reminiscent of the ornamental bird-of-paradise, topped by a cluster of long leaves that become easily divided along their parallel veins. Although bird-of-paradise is in a sister plant family, the plant has a floral structure somewhat similar to the banana as well. But bird-of-paradise bears a dry fruit that opens to release its seeds, while wild bananas bear a fleshy fruit that doesn't. A multi-seeded fleshy fruit is defined botanically as a berry. You'll recognize that our old friend the tomato is also a berry. The big difference between berries created by wild bananas and those created by domesticated bananas are the seeds in the former—and the lack thereof in the latter. Pollination and fertilization are necessary for proper fruit development in wild bananas. The sweet-acid starchy pulp of the wild fruit is thick with black seeds about the size and hardness of double-O BB shot. The seeds of wild bananas are products of sex. If the female flower remains unpollinated, its gynoecium swells a bit on the branch, persisting as small empty shells. In the case of domesticated bananas, those unpollinated fruits develop on their own, filled with seedless pulp.

The spontaneous development of seedless but otherwise normal and flesh-filled fruit is called parthenocarpy. Other well-known parthenocarpic fruits include pineapples, Washington navel oranges, fuyu persimmons, and some clementine mandarins. Wild banana plants produce parthenocarpic fruits very rarely and only if they are not pollinated, but easily produce seedy ones when they are. Thus, the cultivated banana, which produces only seedless fruits under field conditions, is fully female-sterile. Otherwise we would be shattering our teeth on the hard seeds. Careful coaxing by a banana breeder can persuade some types of domesticated

FIGURE 3.1 Banana plant with inflorescence and male flowers. The pseudo-stem is presented with a longitudinal cutaway to reveal leaf sheaths. The banana flowering stalk is composed of developing fruits and a cluster of developing male flowers in the very tip. The underground view shows a stem with roots and a sucker emerging on the left.

banana plants to yield a few true seeds under pollination, but it is far from easy (Simmonds 1966; Koeppel 2008). Contrary to the popular view, the tiny black specks that populate the central core of the commercial banana are not "seeds." The not-so-urban myth gets it almost right; they are unfertilized, aborted ovules.

Bananas do not grow on trees. A young plant shoots up what is called a pseudostem assembly that grows, producing multi-foot-long leaves, until it flowers. The true stem is underground. The inflorescences start as massive clusters of big leathery purple leaves. When individual clusters first open, they reveal creamy yellow, functionally female flowers at their base. As the bud elongates, floral sexuality briefly transitions through bisexual flowers to the creamy-to-pink-to-red male flowers at the tip (Robinson 1996). (We could call this sequence a slightly leaky kind of protogynous monoecy. Find that last clause obscure? Go back to chapter 2 for a refresher!)

The next time you make a banana split, before you peel that long yellow one, stop to check out its tripartite construction. In contrast to tomato's floral five theme, bananas are all about threes and multiples of three. The slim male flowers and the larger female flowers have three sepals and three petals. The nectar has a high sugar content, attracting an array of floral visitors from honeybees to sunbirds to bats. Male flowers have six stamens (including one that isn't functional, a staminode), while the female flowers feature three fused carpels set beneath the perianth (a configuration known as an inferior ovary) (Robinson 1996). Note that these semi-showy, animal-visited flowers topple a banana superlative, the popular myth that the banana is "the world's tallest grass" (Chapman 2007). Nope, it's in its own family, the tiny Old World tropical Musaceae, with less than

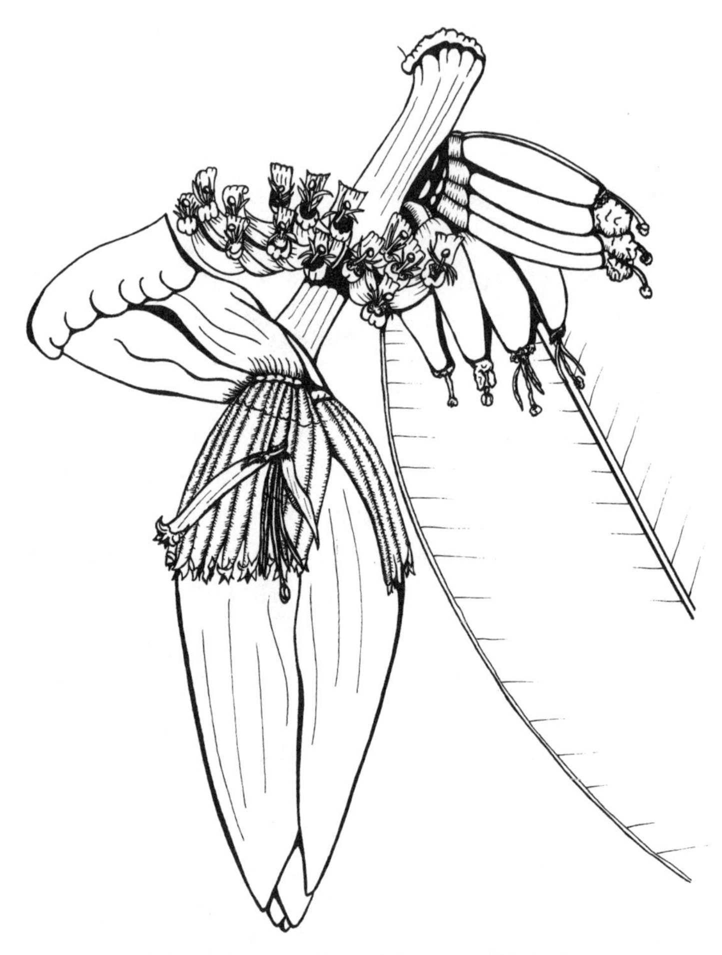

FIGURE 3.2 Male banana flowers and banana inflorescence with developing banana fruits. Central male flower opened to expose the five functional stamens plus one that is not functional (a staminode), which surround the non-functional pistil with three lobed stigmas. One of the stamens has some pollen grains along its margins. The most developed fruits (aka "fingers") show dwindling perianths.

a hundred species. (The world's tallest grass is a bamboo species.)

Wild and cultivated banana pseudostems develop for seven months to over a year until they mature. They flower once, produce fruit, and then die. If a dying pseudostem has sufficient energy and opportunity during its lifetime, it dies surrounded by one or more young pseudostems (see figure 3.1). These "pups" or "suckers" are borne from buds on the true underground stem. The new generation of pseudostems, in turn, may grow to flower and die. If the underground stem spreads successfully over time, it may fragment into physiological discrete—but genetically identical—individuals. Or the plant may be intentionally fragmented by a farmer who replants pseudostems.

You might notice that the concept of "individual" gets a little fuzzy with vegetatively reproducing organisms like bananas (or even some animals, like corals). The idea has stressed out some biologists. John L. Harper, the twentieth century's most influential plant population ecologist, became so frustrated with the conceptual gray zones associated with the terms "vegetative spread," "vegetative growth," "fragmentation," and "vegetative reproduction," he finally decided to avoid the term "vegetative reproduction" and replace it with "vegetative growth": "If a tree spreads vertically we call it growth, but if a clover spreads laterally we call it reproduction—nonsense" (Harper quoted in Turkington 2010). Harper (1977) suggested that other plant ecologists should do the same. That suggestion has not been generally embraced by plant scientists; a Google Scholar search today (December 2017) for the character string "vegetative reproduction" yielded more than sixty-two hundred hits *in the last four years.*

In contrast, Harper's encouragement of using the preexisting terms "ramet" and "genet" for sorting out genetically identical individuals versus those that are genetically different has met with general approval. "Genet" describes the individual or individuals descended from a single instance of sperm-egg fusion. "Ramet" refers to an organismal unit, which may or may not be fully physiologically independent of others of the same genotype (those comprise the genet). In humans, most individuals comprise a single ramet and a single genet. Identical triplets comprise three ramets and one genet. Harper (1977) described the enormity of plant genet variation in nature concisely, "An individual genet may be a tiny seedling or it may be a clone extending in fragments over a kilometer." The contemporary view among most ecologists is that physiologically discrete ramets are ecologically distinct individuals and that a genet describes all of the individuals that share the same genotype—that is, all the ramets that make up a genet would be members of the same clone.

Thus, wild bananas can reproduce both by seed and suckers; cultivated bananas, only by suckers. Reproductively, domesticated banana plants are self-copying machines. A sucker can be moved and replanted, spontaneously growing a new underground stem. Proto-farmers who found an individual whose fruit they liked (maybe tasty and not *so* seedy) could dig up its suckers and replant them in a nearby clearing or even along the trail. In a few years, an industrious proto-farmer might have had reproduced dozens of plants of that genet.

Skip ahead a few thousand years. During the Age of Discovery and Conquest, that model took on global proportions as plant explorers transported a relatively small number of favorite clones out of Southeast Asia and introduced them to the plantation agriculture of tropical colonies. The legacy of the banana diaspora is a few hundred different genets in gardens and farms worldwide: some long, some stubby, some red, some yellow, a rainbow of flavors. With the emergence of the twentieth century, the confluence of the Industrial Revolution with plantation agriculture led to the propagation of a single globally favored banana genet for export from the tropics to waiting markets in the industrial north. Foreshadowing the uniform and reliable vehicle of Henry Ford, the banana industry had found the uniform and reliable banana clone. Everyone was eating the Gros Michel banana (aka Big Mike).

Reliable Uniformity

If you ask any long-term restaurateur, you will find that the secret to getting customers to come back again and again depends on making their favorite meal exactly the same again and again (Bourdain 2000). Constant change—for example, a port-cherry sauce on the duck this week and a lime-dill sauce the next—will send patrons scurrying. If the newspaper food critic's blog tells you to try the tempeh, you won't be pleased to hear that it was replaced by tofu. Part of fast food's global success is that you pretty much know what you are getting. A McDonald's burger and fries meal hardly varies from Riverside, California, to Elk Grove, Illinois, to Uppsala, Sweden. (Actually, McDonald's fries in Sweden are a pleasant surprise.) We are especially dismayed by nasty shocks such as the apricot that never ripens properly or the clementine mandarin with a dozen seeds.

From farmer to grocer, the whole modern industrial food

production and delivery stream depends on predictable uniformity. For field crops, seeds must germinate at the same time, and plants must grow uniformly. Grain must mature uniformly so that a farmhand swinging a scythe or driving a combine can harvest a field in a single sweep. For horticultural crops, packing lines work best when fruits have predictable shapes and sizes to be sorted into standardized boxes. Retailers don't like it when some fruits arrive ripe, others are immature, and the rest are rotten. Given a choice, most customers will leave the oddly shaped tomato or the unevenly colored orange in the bin.

The major sources of biological variation are environmental differences and genetic differences. First a few words about environmental variation. Dips in the field here and there can result in slightly wetter or drier patches. A spray rig that dispenses pesticide or fertilizer might double-spray the crop plants at the end of a row as it makes its U-turn to go down the next. During vacations of my youth, I picked through the fossil-rich rocks of the stone ridges along farm field edges in Wisconsin's Door Peninsula. Years later, I learned that nineteenth-century farmers created those piles as they removed frost-heaved stones to make their fields more uniform and easier to manage.

Environmental variation occurs on a coarser scale as well. Citrus, avocado, and other tree crops thrive in the foothills of California's mountains from the Mexican border and north into the Sacramento Valley. Along that transect, they experience different regimes of temperature, rain, soil type, and time in the sun. Persimmon trees on a north-facing slope near the Mexican border in coastal San Diego County just don't behave the same as those on a west-facing slope 450 miles north in Tulare County's Sierra foothills. Decades of weather data and experience with their trees have guided grove managers to predict the quality and quantity of their crop and when it will be ready to harvest relative to other growers in the state. Industrial farmers have become increasingly sophisticated in dealing with environmental variation. The cutting edge known as precision agriculture enables a farmer to treat every plant as uniformly as possible. Combining precise spatial information from a Global Positioning System (GPS) with high-tech field machinery can be used to alter the external environment (flatten that field). Add soil and plant sensors for nutrient and water status, mix with Internet cloud-based data analytic tools, and you've got a recipe for undoing some of Mother Nature's environmental tweaks.

With these modern tools, crop, land, and irrigation management can attain a refined level of environmental uniformity. But there's a limit to controlling for uniformity if a crop is genetically diverse and that diversity is manifest over plants and their products. Genetically variable outcrossed sexual seeds throw off a tremendous amount of variation. We enjoy the kaleidoscope of colored kernels on an ear of dried "Indian corn" for a Thanksgiving decoration. But a kaleidoscope of interplant variation is a nightmare for those seeking to make their product as predictable and uniform as possible. However, not just any single genet will do. Some genotypes are developmentally unstable. They simply won't be uniform, no matter how nicely you treat them. Others might be uniform within any given locality, but vary among localities—sensitive to what might seem minor variations in terroir. From the point of view of a global banana industry, the very same banana (i.e., in the eyes of customers as well as those who pack, move, and sell the product) must be produced in Ecuador, Philippines, Guatemala, Costa Rica, and Colombia.

For the commercial banana industry and trade, the Gros Michel genet must have seemed like Mary Poppins, "practically perfect in every way." It was a tasty tropical treat for nineteenth-century urban customers who enjoyed their fresh fruit in the summer but had to turn to canned or dried fruit in the winter. With a thick skin, it traveled well. More importantly, the variety grew and yielded well and uniformly in a variety of tropical countries and continents. After harvest, it survived the long boat ride to cooler climes, delivering a good banana for the first half of the twentieth century. Peter Chapman (2007) explains:

> United Fruit [Company] was a pioneer of mass production. With its one-size-fits-all banana, the company beat Henry Ford, the man often credited as the pioneer of standardization, by a number of years. Big Mike was on the shelves at the turn of the 20th century; the Model T, on the other hand, rolled off the production line in 1908. . . . United Fruit's bananas were the forerunners of those products we know today: the cup of cross-cultural coffee foam; the multinational hamburger.

Because most important crops reproduce only by sexual seed, they cannot be clonally propagated. Not surprisingly, the genetic variation generated by sexual reproduction is an obstacle for many folks looking to deliver a better crop product. For the past quarter century, some plant biotechnologists have argued that future crops should follow the banana, dispensing with sex entirely. Specifically, they are titillated by the idea of varieties that replicate the maternal plant via reliably uniform, asexual, apomictic seed (e.g., Hand and Koltunow 2014). One proposal is that the plant breeders would maintain sexually fertile lineages that, when crossed, would create apomictic offspring. A second approach would be to genetically engineer plants to be apomictic. The seeds produced by either method could be delivered to farmers, who benefit from the crop's uniformity.

Depending on whose prose you read, asexual seed could be a real boon to farmers—or not (National Research Council 2004). Some scientists see social justice in apomictic seed (Jefferson 1994). Presently, the vast majority of seed of outcrossed crops are so-called hybrid varieties. Hybrid varieties are high producing and highly uniform, resulting from crosses between two highly uniform inbred varieties. (Note: Commercial hybrid varieties don't have anything to do with hybrids between species. More on hybrid varieties in chapter 5.) But when these plants from hybrid seed mate with each other, sex scrambles their genetic structure apart. The offspring are both variable and inferior. Decades ago, I visited a Mennonite farm near Lancaster, Pennsylvania, and saw struggling dwarf corn plants on the edge of a productive field. The farmer explained that these dwarfs were the sexually mixed-up seed that had fallen into the soil from the prior year's hybrid crop. Despite receiving the same rainfall, weed removal, and fertilizer, the volunteer plants were less than half the size of the hybrid crop plants growing just inches away.

Farmers can't get a decent crop by saving and replanting the seeds produced from a hybrid variety. If they like what they have and can afford it, farm operators must return to buy more from the seed company every year. Farmers in poverty are at a disadvantage. If they can ever afford to buy hybrid seeds once, their onetime gains would still be insufficient to help them to buy new seeds every year. With apomictic varieties, farmers wouldn't have to buy seeds year after year because their hybrid crops would produce progeny identical to and as vigorous as their mother. Proponents for this scenario are obviously in the public research domain—for example, nonprofits and governments. A prominent example is the independent, nonprofit Cambia (www.cambia.org), an institute that is exploring how apomictic varieties can be created by genetic engineering and more advanced molecular technologies.

But given the cost and difficulty of engineering apomixis, there is the concern that corporations that could bear the cost might create and use apomixis technology for their own profit-oriented purposes, and that poor farmers might not be able to access the expensive products. There is also the question of whether asexual seed is always a boon. A farmer who buys a sexually sterile apomictic variety has no opportunity to engage in his own crop improvement. That's especially true for farmers in developing countries who don't routinely use hybrid varieties. Instead they grow open-pollinated landraces that evolved constant experimentation. The most experimental farmers grow their crops as a mixture of plants descended from saved, exchanged, and even purchased seed, often actively or passively participating in activities that increasingly adapt their crops to local conditions. Concern about corporate control of apomixis led to the 1998 "Bellagio Apomixis Declaration," urging "widespread adoption of the principle of broad and equitable access to plant biotechnologies, especially apomixis technology," and encouraging "the development of novel approaches for technology generation, patenting, and licensing that can achieve this goal" (Group of Reproductive Development and Apomixis 1998). At the moment, plants genetically engineered to create apomictic seeds are lab-bound, a long way from the field and even a longer way from distribution to farmers.

One thing that everyone agrees on is that a clonal crop is genetically very uniform. That's both the good news and the bad news because, despite its tremendous popularity, the Gros Michel is nearly gone. Not because a better-tasting or a better-yielding banana came along but because, according to banana geneticist and crop evolutionist N. W. Simmonds (1995), "bananas constitute one of the best examples in the history of agriculture of the pathological perils of monoclone culture."

Life without sex can have its downsides.

"What use is sex?" "What good is sex?" "Why sex?" "Why have sex?" Are these complaints of a frustrated celibate? No, they are titles of scholarly academic papers dating from the last half of the twentieth century and continuing to the present day (Maynard Smith 1971; Michod 1997; Wuethrich 1998; Otto and Gerstein 2006). While sex may be a bit of a mystery to us all, it has been the grand enigma of evolutionary biology for decades.

Sex is nearly ubiquitous, but it hasn't been easy to explain its ubiquity. Reproduction by sex is universal for humans, mammals, and birds, and nearly universal for fishes, amphibians, reptiles, and most other animal species. A million, probably more, animal species can or must reproduce by sex. A sizable minority of animals have the option of both sexuality and asexuality—for example, some corals, water fleas, and aphids. A small fraction is strictly asexual, from some rotifers to certain wasps and a handful of lizards. No large group of animals is sex-free. The same is true for other organisms. Of the more than quarter million angiosperm species, many more than half are sexual with the option to reproduce asexually. But flowering plant species that have wholly dispensed with sex are few and far between. The ability to reproduce *both* with and without sex is very common for primitive plant species, such as ferns, mosses, and horsetails. But even for these groups, species that never create functioning male and female parts are infrequent at best.

It is a habit of evolutionary biologists to explain the persistent and widespread properties of organisms with a bit of Darwinian logic. Prominent characteristics are readily explainable as adaptations. You find plants that conserve water in the desert. Animals with spots survive and flourish in habitats dappled with light and shadow. Light-skinned folk originate from cloudy areas. When they live in sunny climes, they suffer sunburn, skin cancer, and vitamin D poisoning. Darker-skinned folk who tend to be native to regions with intense sunshine are far less vulnerable to these problems. Sexual reproduction is a common feature of all major groups of organisms because . . . um . . . because . . . well . . . hmmm?

The explanation for sex isn't straightforward. Sex is a hassle. To reproduce without sex, an organism can dedicate a cell toward creating a new individual, pump it up with some resources, and eventually set its baby free. The organism that uses sex to reproduce has a greater challenge; it has to create gametes that have to find other gametes. The process of seeking or attracting those other gametes typically involves allocation of resources to special structures and, in the case of animals, allocation of resources and time to special behaviors.

To evolutionary biologists, sexuality involves an additional hassle. Sex appears to violate the rule that an individual bearing a Darwinian adaptation cannot be successful unless it leaves more genes than the alternate trait. (According to Darwin, "for the good of the species" is not good enough.) For sexual reproduction to serve the evolutionary benefit of an individual, organisms engaging in sex should leave more of their genes to the next generation than those that reproduce without sex (Williams 1975). The alternate trait is asexuality, and that should leave more genes than sex.

The paradox is explained by simple arithmetic. An individual that creates a seed by apomixis passes on all, 100%, of its genes. An individual that creates a sexual egg puts half, 50%, of its genes in that basket. The fertilizing sperm from another individual makes up the other 50%. The sexual seed passes on half as many of its maternal genes than does an asexual seed (or any other asexually produced offspring). Evolutionary biologists did their arithmetic and gasped. An individual that procreates asexually passes on twice as many of its genes per offspring as the one that procreates sexually. Dispensing with sex yields an immediate twofold benefit. A similar, but more complicated, mathematical argument comes to the same conclusion for organisms that have the option of self-fertilization. For sexual reproduction to be maintained in a species, its benefit must be *huge* and obvious. A myriad of theories have been developed—and continue to be developed—suggesting what that "obvious" benefit might be. Most proceed from the simple fact that sexually produced offspring are genetically different from each other and from their parents and ask the question "What's the advantage to being different?" The theories can be crudely lumped into three groups.

One approach, nicknamed the Lottery model, says that

the asexual parent is analogous to an individual who buys a pile of lottery tickets, but each with the same number. The sexual parent takes the alternative strategy of buying a pile of tickets that are all different from each other. If the winning conditions are very predictable, that is, the environment for the progeny is identical to that of the parent; the asexual parent is the *big* winner. But if environments change unpredictably in time and space, the sexual parent has a better chance of holding the winning ticket. Notice that this argument works well for those individuals who have lots of progeny (tickets) and those progeny end up in unpredictable environments. It works especially well if progeny end up dispersed away from the maternal environment.

The next model requires that siblings interact and compete over some set of limited resources. The Elbow Room model proceeds as follows: If all the progeny are identical, they all have the same needs and specialized adaptations. In those organisms that have limited dispersal, asexual progeny are likely to hurt each other by equally depleting the same critical resources. In contrast, sexually produced progeny have somewhat different resource adaptations and needs from each other and will probably interfere less with each other as much as those that are identical. Three mechanically endowed sisters in a small and remote west Nebraska town are less likely to prosper setting up competing Ferrari repair shops compared to three siblings diversely talented as butcher, baker, and candlestick maker.

The final model starts with the premise that the world is full of biological enemies: predators, parasites, and disease organisms. But biological enemies do not obtain their nutrition willy-nilly. These organisms have specializations for finding and attacking their prey and hosts. Under these circumstances, it is good to be different from others of your species and, if dispersal is low, from your siblings. It's good to be the rare type in the "eyes" of the enemy, that is, inconspicuous. Let's consider a crude example that involves a predator. A bird that is visually attuned for attacking bright blue beetles might miss those few that are dull green, brown, et cetera. An asexual bright blue beetle and its kids are doomed, but a sexual blue beetle has the opportunity for sex to shuffle the genes with the evolution of novelty every generation. Under strong evolutionary pressure from a biological enemy, the rare type becomes the common type. Sex might be the trick to provide temporary escape for some progeny in the next generation, as natural selection begins to favor the birds that are visually attuned for dull colors

Disease organisms—worms, bacteria, fungi, and other simple organisms—not predators, are thought to run the show. They have short generation times (as rapid as once every twenty minutes) and thus can come in huge numbers (bazillions or more). While most have the option of sex, many get their variation from high rates of mutation (spontaneous genetic change). That combination allows for blazingly fast evolution by natural selection. They evolve to pursue and mow down the most common susceptible host genotype (or set of similar genotypes). The once-rare host type becomes common. Then selection sorts among the variation in the now-beleaguered disease organism. The once-rare-but-now-common host gets hit with the newly evolved disease organism. And so the evolutionary cycle continues.

A sexually evolving host cannot rest (and certainly cannot revert to asexuality) when pursued by rapidly evolving disease-causing microbes. In this co-evolutionary race, both the disease organism and host are running as fast as they can to stay in one place. And that's why the theory is referred to as the Red Queen model, after a broader co-evolutionary theory (the Red Queen hypothesis) (Van Valen 1973). The theory's name pays homage to the Red Queen's race of the same type in Lewis Carroll's (1871) *Through the Looking-Glass*, when she says:

> "Now, here, you see, it takes all the running you can do, to keep in the same place. If you want to get somewhere else, you must run at least twice as fast as that!"

A massive body of experimental and descriptive scrutiny of these major models, their variants, and some other hypothesis has accumulated over the last handful of decades. Lively and Morran (2014) have thoughtfully reviewed the state of the art. While the definitive answer may not yet have arrived, a consensus is certainly emerging. The Elbow Room model appears to be the big loser. Several experimental studies have been conducted involving competition between sets of asexual and sexual progeny from the same mothers. They rarely show an advantage to the genetically more diverse sexual offspring (e.g., Ellstrand and Antonovics 1985); at best, whether increasing genetic diversity reduces competition appears to be idiosyncratic and hardly universal (File, Murphy, and Dudley 2011). The Lottery model appears to have some validity in the minority of cases examined, especially for those organisms that create vast numbers of progeny that are widely dispersed (Antonovics and Ellstrand 1985). With tremendous support from studies of natural populations and strong support from experimental studies, the Red Queen is the big winner. Data from both plants and animals reveal that there's an advantage to being the rare genotype in the presence of a biological enemy. Many of the studies that actually measured the evolutionary benefit, in terms of number of genes represented in the next population, have shown that sex's advantage—in the face of biological enemies—often exceeds the twofold evolutionary cost of sex.

It turns out that the Red Queen has something to say about bananas.

Lest we forget: "Bananas constitute one of the best examples in the history of agriculture of the pathological perils of monoclone culture" (Simmonds 1995). The Red Queen is behind the demise of the Gros Michel. Once you have huge swaths of extreme biological uniformity, it is only a matter of time before some pest stumbles upon that sea of free reliable and uniform nourishment. In the case of the Gros Michel clone, the culprits were two fungal species—one that caused Panama disease (even though it was first recorded in 1876 in Australia) and another that caused a disease called yellow Sigatoka. It turned out that the billions of Gros Michel plants were fully susceptible to both of these as well as an array of minor pests. The Panama disease organism was—and continues to be—particularly nasty. No control methods have yet been found to save infected susceptible banana genotypes. Panama disease destroyed most of the Gros Michel plantations in the Americas in the first half of the twentieth century (Stover and Simmonds 1987). Faced with rapidly disappearing bananas, the only other solution was to either find or build a better banana. The quest began in earnest in the 1920s (Robinson 1996). That was the same time that "Yes! We have no bananas" became popular. But whether the impending demise of Big Mike was the inspiration for the song remains a matter of speculation.

As the Gros Michel lurched to near-extinction, the fruits of those research programs began to be planted. Gros Michel was replaced with a set of very closely related resistant genets (all, in fact, are clonally descendants of a single genotype, all mutants, no sex involved) known as the Cavendish subgroup. Cavendish was a preexisting clone that wasn't susceptible to the nemeses of Gros Michel. Today, the Cavendish subgroup rules. Roughly half of the world's nine million acres of bananas belong to this group. Furthermore, Cavendish types are the banana of intercontinental trade. Currently, cultivation of Gros Michel hangs on in parts of the world where those nasty fungi haven't yet found it (Robinson 1996; Koeppel 2008).

But the Red Queen's race doesn't stop. Organisms that stand still, like the billions of Cavendish plants, will eventually fall behind. As Cavendish began to rise to dominance, new disease organisms started attacking it. The worst contemporary banana disease is black Sigatoka, caused by a close relative of the yellow Sigatoka fungus (Robinson 1996).

More worrisome in the long run is an old friend. The original Cavendish-resistant Panama disease culprit has now been named Race 1. Panama disease fungi have evolved; the Race 1 genotypes are being replaced by one known as Race 4, which first appeared in 1965. Cavendish has no resistance to Race 4. The evolutionarily new and improved Panama disease organism has wiped out thousands of Cavendish acres in Southeast Asia. Since then it has been identified in the Pacific, Australia, Africa, and the Middle East (Ordonez et al. 2015). Worse yet, in 2011 Cavendish bananas in India started succumbing to what appears to be a new genetic variant of Race 1 (Thangavelu and Mustaffa 2010). The bad news is that Cavendish is fully sexually sterile.

The good news is that many of the banana breeding and screening programs created to replace Gros Michel are still in place. Again, alternate varieties are being sought, bred from clones that retain a bit of seed fertility or are engineered (Koeppel 2008). For our food crops that are largely asexually propagated, plant breeders at private companies, universities, and other public institutions get to play the crucial role that periodic sex does in natural populations. Worldwide, promising disease-resistant clones are being created, both by traditional methods and by genetic engineering. But before a promising clone is multiplied by the thousands, it must be field-tested and consumer-tested. Is this banana's second race against time? If Race 4 fungus is accidentally introduced to the commercial banana heartland in the tropical Americas before breeders find an alternate genotype, we could yet again be singing, "Yes! We have no bananas."

The sort-of-good news is that Cavendish is not the predominant clone in areas where the banana is a staple. The hundreds of millions in the tropical world that depend on bananas for their sustenance depend on different clones. Often, even within a given region, multiple clones are grown. But imagine a situation in which millions of humans within a single region were nutritionally dependent on some food from a single locally grown clone.

Scary, indeed.

Ireland's infamous potato blight and ensuing famine is illustrative. Potato plants rarely flower and set seed even less frequently. The crop is typically maintained by cutting up potato tubers into pieces that are individually planted. A potato tuber has the botanical name of rhizome, labeling it as an underground horizontal stem. Each "eye" of the potato tuber is a bud with the potential to grow into a new plant with the exact genetic makeup as the parent tuber and its clonal siblings. In the early nineteenth century, Ireland enjoyed a population explosion. The surge was fueled by the easy-to-grow and calorie-dense potato. The tuber had become a staple for at least half the population. In contrast with the dizzying genetic diversity of the spud in its Andean homeland, where tubers come in a kaleidoscopic array of genets with different colors, shapes, sizes, and textures, Ireland's historic crop was based on a single widespread clone, Lumper. Lumper spread and flourished, waiting for a disease organism for which it had no immunity. When the late blight water mold arrived, potato plants started melting with rot in the fields, and the tubers did the same in storage. Each year's harvest was worse than the prior. The Great Famine from 1845 to 1852 was catastrophic. Well over one million Irish died from famine, an equal number emigrated, and presently, more than a century and a half later, the population has still not recovered to its historic high (Ristaino 2002). Lumper rapidly approached extinction and would have met that fate if not for an Irish farmer who is maintaining it for its historic value (Zuckerman 2013).

Over the time that life has resided on our planet, the total number of species has gradually increased, despite the occasional mass extinction (Bennett 2013). That means that speciation rates generally exceed extinction rates. By the best available scientific reckoning, we live in the era when the number of species peaked, within the last 20,000 years or so. In fact, new species continue to evolve spontaneously within historic time, even new food plant species. Bread wheat is a good example (Feldman, Lupton, and Miller 1995). Sex was the engine that mixed and matched the genes to create most of those species. Lineages that practice sexuality are thought to be much more successful at sustainably diversifying over time compared to those that are strictly asexual. Those arguments generally depend on the assumption that speciation rates in sexual lineages exceed those of asexual lineages (Barraclough, Birky, and Burt 2003). In fact, although few, if any, strictly sexual species have known asexual parents, many strictly asexual species are known to have recent sexual ancestors (the female-only whiptail lizard species are a nice example [Sites et al. 1990]).

But let's make it clear that neither sex nor gigantic population sizes nor human intervention—separately or in concert—is always capable of preventing extinction altogether. In fact, human activities are thought to account for the significant global decline in the number of species in the past few thousand years (when was the last time you saw a giant ground sloth?) (Barnosky et al. 2011). We must depart from the plant world for a compelling, well-documented example.

As recently as 175 years ago, the most numerous sizable vertebrate in the world was the passenger pigeon. The bird was North America's largest member of the dove family, about a third bigger than the mourning dove. The total population size of the passenger pigeon during early European colonization was perhaps five billion. That's roughly ten birds for each human alive on Earth at that time. The average size of a nesting colony was figured to be thirty-one square miles. The passenger pigeon's numbers increased as the colonists put an increasing number acres to the plow, affording the birds with an abundant source of grain to supplement their regular diet of insects, berries, and particularly nuts from the nearly continuous beech, maple, chestnut, hickory, and oak forest of the eastern side of the continent (Cokinos

2000; Forbush 1936). That forest was home to their roosting sites and nesting colonies, with the occasional tree collapsing under the mass of tons of visiting birds. Pioneering early nineteenth-century American naturalist Alexander Wilson estimated one flock to contain well over two billion birds. Author Christopher Cokinos (2000) brings that number into context:

> Assuming each pigeon was about 16 inches long, a line of 2,230,272,000 Passenger Pigeons would have equaled 35 billion inches or 3 billion feet. That's 563,200 *miles* of Passenger Pigeons. In other words if Wilson's flock had flown beak-to-tail in a single file the birds would have stretched around the earth's equatorial circumference *22.6 times*.

Despite these once fabulously large numbers, the species is now extinct. Pioneers and market-gunners harvested thousands of birds (sometimes thousands at a single time) for food for themselves, their farm animals, and the market. The birds' abundance invited recreational slaughters. Equally important, rapid removal of eastern forests eliminated passenger pigeon nesting sites and natural food sources. The last passenger pigeon, lonely Martha, died in the Cincinnati Zoo in 1914 (Cokinos 2000; Forbush 1936). Obligately sexual (asexuality in birds is extraordinarily rare, except for—and why is this *not* surprising?—the domesticated turkey [Tomar et al., 2015]). The passenger pigeon didn't win the Red Queen's race against human predators. Furthermore, when human efforts to save that species began in earnest in the 1870s, they proved futile. It's hard to say whether the sophisticated science of the early twenty-first century could do any better (but that doesn't mean that there aren't efforts to bring those guys back to life [cf. Sherkow and Greely 2013]). Sex can do a lot to save a species, but, regrettably, it is not an evolutionary panacea.

Nonetheless, sex is good enough to keep life going overall. There's not a successful major evolutionary lineage that has dispensed with sex altogether. Can a few billion years of evolution be wrong? In her 2008 novel, *Benny and Shrimp*, Katarina Mazetti observes, "'Love' is how a species answers the need for genetic variation; otherwise you could easily just taking cuttings from the females." That's true enough for bananas and potatoes in the face of diseases. But it can cause other problems as well. The next chapter explores how the romantic life of the avocado is compromised when whole populations of trees are created by "taking cuttings from the females."

Yes, We Do Have Bananas! Puffed Pancake

It seems ironic that a phallic fruit should be the poster child for celibacy given that it is as sterile as a rock. But so it goes. As you cut into the bananas for this treat, check out the aborted pre-seeds to remind you that life without sex presents a precarious evolutionary future. And count your blessings, because, yes—at the moment—we do have bananas.

I've been making this adapted "Dutch Baby" on Sunday mornings and for visitors for decades. Originally from an apple pancake recipe in a catalog of a now-defunct cookware company, I swapped out the apples for bananas and cut the butter and sugar by about two-thirds. To my taste, the caramelized bananas work well with the suggested spice mix.

But you can go crazy with your spice shelf. Of course, there's nothing wrong with adding other fruits or even nuts (how about pecan pieces?) to the bananas.

2 or 3 ripe to overripe bananas, peeled and sliced
3 tablespoons butter
3 large eggs
½ cup milk
½ cup flour
⅓ cup melted butter
⅓ cup sugar mixed with 1 tablespoon cinnamon, ½ teaspoon cloves, and ½ teaspoon ginger (a pinch or more of cardamom might be fun)
Lime quarters, optional

Serves two hungry folks, or four, if supplemented by plenty of breakfast side dishes.

It's a good idea to read this recipe through in its entirety before executing it. The timing is a bit tricky.

Preheat oven to 475°. While oven is warming up, melt three tablespoons butter in a ten- to twelve-inch ovenproof skillet. When butter is melted, spread it around to coat the inner surface of the pan to prevent subsequent sticking of fruit or batter. Add banana slices and sauté until softened and slightly caramelized. (I stick the pan in the oven and let the bananas sauté while I make the batter!) Use a beater or blender to create a frothy mixture of eggs, flour, and milk. Pour the egg mixture over the sautéed bananas. Place the skillet into the oven. Bake about twelve minutes until the pancake browns on the edges. Then remove it briefly from the oven. Pour the melted butter over the pancake, and sprinkle the spice sugar mixture on top. Be careful but work expeditiously so that you can return the pancake to the oven before it cools. Once it is in the oven, bake for six minutes or until the sugar melts. It should be puffed up soufflé-style, but will not hold that profile long. Slice and serve. If desired, squeeze lime over slices.

Unexpected visitors? Need a quick treat? Reheated leftovers (ha!) work well with a scoop of vanilla ice cream.

4 *Avocado*

TIMING IS EVERYTHING

The avocado is a food without rival among the fruits, the veritable fruit of paradise.
—David Fairchild, plant explorer

For lovers of the avocado, the nutty flavor and the firm and yielding creamy gold-green flesh create an urge to poetic praise. Botanically a fruit; culinarily a vegetable. But for avocado zealots, it *can* act as a fruit in the culinary sense: any of the standard varieties can be used to make ice cream, and certain varieties can serve as a dessert fruit. The fruit might not be as overtly sexy as a banana. But that didn't prevent the Aztecs from naming the "ahuaca-tl" fruits they discovered during their imperial expansion with the same word they used for "testicle" (Karttunen 1992).

But how to sell Fairchild's "fruit of paradise" in diet-conscious modern times? After all, a good-sized fruit has about 300 calories, more than 80% of which are fat calories. The California Avocado Commission's answer was the sometimes blonde and perennially leggy TV/film star Angie Dickinson. In her 1982 TV ad, Dickinson asks, "Would this body lie to you?" and then recounts the avocado's vitamin- and mineral-based virtues.[1] The photo print version, suitable for hanging in the men's dorm, measures the avocado's calories with a "per slice" value of 17. Surprisingly, weight for weight, the avocado is nutritionally superior to many other high-fat foods. An ounce of low-sodium avocado has about 25% of the calories of one ounce of butter while packing two grams of fiber (Dreher and Davenport 2013). The fact that the avocado's fat is 87% unsaturated with no cholesterol might make you consider reaching for an avocado instead of the butter dish the next time you make toast (see this chapter's recipe). Whether you are an avocado fan or not, you'll agree that the California Avocado Commission's commercial extended the Aztec tradition of linking the avocado and sex.

But the story of the precarious sex life of the avocado is better suited for the movies than television. The story has subtle parallels to Rob Reiner's 1989 movie *When Harry Met Sally*. In both, timing plays a key role with regard to the importance of finding the right mate at the right time. Biological timing is equally important for the avocado's sexual fulfillment. In fact, timing plays a total of three different roles in getting that good avocado to your plate.

The sexual expression of the avocado's flowers defines its romantic dilemma. But first, let's set the stage with a little reproductive botany. Except for its single-carpel gynoecium, avocado flowers are structured according to multiples of three. That three-ness of avocado flowers is standard for its family, the Lauraceae. The fertilized avocado carpel develops into a fleshy fruit, but it doesn't join the banana and tomato as "berries." Avocado fruits are one-seeded.

1 Still available online: www.youtube.com/watch?v=9288uol1lwQ.

Botanically that makes them drupes (pronounced "droops"). Familiar drupes include peaches, mangoes, and olives (all in different families).

Lauraceae is composed of about three thousand mostly tropical trees and shrubs (Heywood et al. 2007). Except for the avocado (*Persea americana*), the family's few contributions to the kitchen are essential to the cook. You might not have an avocado at home, but I bet if you make pasta sauce and occasionally bake pumpkin pie, your spice collection includes bay leaves (*Laurus nobilis*) for the first and cinnamon for the second (either mellow *Cinnamomum verum* or, more likely, the robust "hot" *Cinnamomum cassia*, aka cassia) (McGee 2004).

The flowers of the Lauraceae are small but not tiny. Walking through a flowering avocado grove in spring is a subtle experience. Avocado flowers' scent has yet to inspire a musician to write "The Avocado Blossom Special." Avocado flowers are about four-tenths of an inch in diameter. Relatively modest compared to the tomato and banana flowers that we've met, they are still showier than those of most plant species. The three greenish-white sepals are nearly identical in size, shape, and color to the three petals, giving botanists yet another opportunity to create a name. The parts of a perianth with nearly identical calyx and corolla don't need to be called petals or sepals. Collectively, they are called tepals. Go figure. Depending on the avocado variety, a tree can start flowering as early as fall or as late as mid-spring. But it's not the season of flowering that compels the romantic drama. It's how—and when—the actors make their moves. The choreography includes both the flowers themselves and their pollinators.

Avocado flowers are bisexual but dichogamous (Bergh

1973; Salazar-García, Garner, and Lovatt 2013). Specifically, individual flowers are protogynous, which, you may recall from chapter 2, means that when bisexual flowers first open, they express only female function. An avocado flower opens twice, each time for several hours. The first time, the greenish gynoecium is erect and receptive, the immature stamens are flaccid, and the flower is functionally female. The tepals close. More hours pass. At the second opening, the whole flower is 10 percent larger. The style of the gynoecium is spent and no longer receptive. But now the nine mature stamens (an inner whorl of three, and an outer whorl of six) are erect and releasing their pollen. The flower is functionally male.

Now for the tricky part.

All the flowers on a tree are in sexual synchrony. That is, when one flower is in female phase on a tree, every open flower on that tree is in female phase. When the flowers reopen as male, all of the tree's open flowers are male. The entire tree is physiologically choreographed to alternate between female and male. You could say that entire avocado trees are dichogamous (Bergh 1973; Salazar-García, Garner, and Lovatt 2013).

With regard to flowering, avocado trees come in two types, not female and male, but A and B. Genetically determined type A and type B trees have male function and female function, but at complementary times of the day. Flowers on type A trees have a cycle of roughly thirty-six hours. An A tree starts the morning with all of its freshly opened flowers behaving as females. Let's follow this particular cohort through its short-lived existence. As the sun rises and midday approaches, the flowers close. They remain closed for twenty-four hours. Another cohort of female flowers opens

FIGURE 4.1 Avocado flowering and flower phases. Early spring blooming with reproductive buds at the ends of newly developed shoots (*center*). Detailed views of avocado flower showing male phase (*left*) and female phase (*right*). Each flower has six tepals and is approximately six millimeters long (see also figure 4.2). Male phase flowers (*left*) have an erect style with a stigma that is receptive to pollen. The inner whorl of three staminodes produces nectar. The anthers have not yet opened to release pollen. Female phase flowers (*right*) open with all nine stamens erect and releasing pollen, an inner whorl of three and an outer whorl of six. The stigma is no longer receptive to pollen.

the next morning of the second day, but the cohort we are following remains closed at that time. In the afternoon of the second day, the first group reopens as males (while the second cohort is napping). By evening these flowers, pollinated or not, are no longer sexually active. The morning of the third day, a third crop of female flowers starts the cycle afresh.

Type B trees have flowers that complete their cycle in less than twenty hours. Each new set of female-phase flowers always opens in the afternoon. They close as evening approaches. Morning sees them reopen as males. By noon

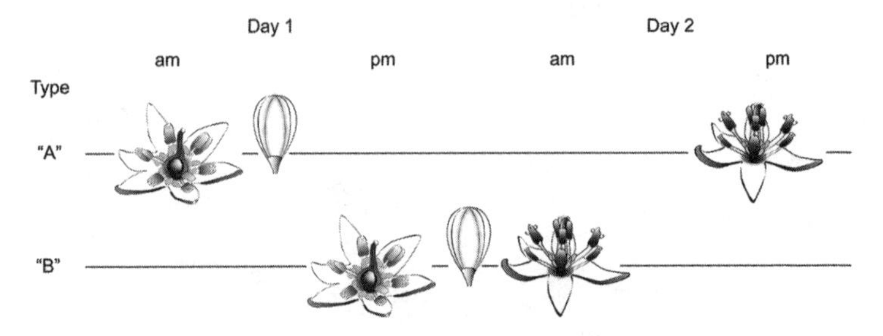

FIGURE 4.2 Timeline of the sex expression type A versus type B avocado trees.

they are no longer dispensing pollen. That afternoon a second cohort of female flowers opens. Thus, when type B trees are functioning as females, type A trees are functioning as males—and vice versa (Stout 1923; Bergh 1973; Salazar-García, Garner, and Lovatt 2013).

Let's personalize this way of life. Imagine waking up every morning as one gender, taking a nap in the afternoon, and reawakening as the other gender. Then imagine it happening every day. Finally, imagine that this switcheroo is standard for everybody, with the proviso that half of the population is male in the morning and the other half is female. This kind of arrangement could give new meaning to some of those country songs of late-night barroom romance. In the world of flowering plants, this kind of sex switching is rare. Whole-plant alternation of female to male to female to male, et cetera, on a daily basis is extraordinary and known from a handful of species. Only a few other food plants have similar (but not identical) simultaneous whole-tree sex switching, including some of the tropical custard apples (in the Annonaceae family, sitting nearby Lauraceae in the angiosperm family tree) and jujube (aka Chinese date, in a very distant family) (Lloyd and Webb 1986).

Avocado trees go out of their way to avoid self-pollination; first, promoting outcrossing by within-flower protogyny, and then by whole-tree protogyny. Still, we know that avocado trees are self-compatible (Sedgley 1979). Pollen can be harvested and its viability maintained under controlled conditions for successful hand-pollination of the next cohort of female flowers. But recall Darwin's (1876b) dictum, "Nature abhors perpetual self-fertilization." How well does sex-switching work to prevent self-fertilization? How rigid is the system? After all, mature avocado trees can produce a million flowers (Bergh 1992). If a developmental wobble happens, an avocado tree can potentially self-fertilize by two different methods, pollination within individual flowers (autogamy) or transfer of pollen between its flowers (geitonogamy).

Autogamy is rare because of the dance of the avocado's sex parts. Students of the avocado have noted that insects occasionally visit closed flowers and open them to get to pollen or nectar (e.g., Ish-am and Eiskowich 1993). If the pollen is viable at that time, such disturbance could aid in self-pollination. Despite abundant research on the avocado flower and its pollination, I am not aware of any study that has shown conclusively that autogamy contributes in any appreciable way to avocado fruit set.

In most regions, a pollen vector is required for avocado fruit set. The general consensus—at least for subtropical regions such as California, Chile, Spain, and the Mexican state of Michoacán—is that insects are the natural pollinators of avocado trees. The following early simple experiment in California describes the compelling need for insect vectors: Whole individual mature trees of two varieties were caged, with and without a hive of honeybees. Those trees without bees averaged no more than five fruits total in the season. Those caged with bees averaged well over one hundred fruits (Peterson 1955). For comparison, a good yield on a mature tree is considered to be three hundred or more fruits per tree.

Autogamy might be important in the tropical avocado-producing regions, such as Florida and the Dominican Republic, but the data remain unclear. Florida avocado researchers used cheesecloth to bag branches with female-phase flowers and those in male phase, excluding animal pollinators too large to find their way through the holes of the cloth. Subsequently, they found that the stigmas of the flowers were frequently full of pollen, especially in the male phase. But they could not determine whether pollination was affected by within-flower pollination or by within-branch, between-flower pollination mediated by wind, gravity, or thrips, teeny crawling insects known to frequent avocado flowers (Davenport et al. 1994). Thrips are so tiny that if you arrange them end on end, a dozen would barely be as long as a honeybee. They are small enough to make their way through cheesecloth, and the researchers observed some thrips on the bagged branches. Did the presence of pollen lead to successful fertilization? Hard to say. The yield of each bagged branch averaged less than a single fruit at maturity.

That's no surprise. Avocado breeders and other scientists had already learned that if they carefully applied viable pollen to female-phase avocado flowers, for *every thousand* flowers pollinated, *less than one* fruit matures (Salazar-García, Garner, and Lovatt 2013). Under natural conditions, when pollen makes its way to a receptive avocado stigma and affects fertilization, the odds are dismal that a mature fruit will result. Many pollinated flowers start developing into fruit, but most fail to complete the journey. Well-pollinated older trees spontaneously drop thousands or even tens of thousands of immature fruitlets, primarily during the first month after pollination. Nasty weather can encourage fruit drop. Long-term cold spells during fruit development are bad. Severe hot spells are even worse. The Fuerte variety, in particular, is so prone to fruit drop that the larger fruitlets are collected for consumers as essentially seedless "cukes," resembling oversize olives or baby cucumbers. Marketed as "cocktail avocados," they are occasionally available in farmers' markets (Salazar-García, Garner, and Lovatt 2013).

Geitonogamy is more likely than autogamy. Avocado researchers have observed a large number, but tiny fraction, of "off stride" flowers on a tree (Stout 1923)—that is, functionally male flowers among an abundance of females, or in female phase when most flowers have switched to male. This irregular pattern is especially true for trees suffering from cold or erratic temperatures (Chanderbali et al. 2013). So *some* within-tree, between-flower self-fertilization is possible, though probably at a low frequency and provided that pollinators are available.

Avocado flowers are known to be frequently visited by many bee species, as well as flies, wasps, beetles, and (as mentioned above) thrips. Subtropical growers often bring honeybee hives into the grove to supplement the other insects that pollinate their trees. The honeybee, domesticated in western Eurasia, appears most effective in pollinating the avocado (Salazar-García, Garner, and Lovatt 2013), even though the tree is native to Mesoamerica and was originally pollinated by diverse wild neotropical bees (Can-Alonso et al. 2005). The honeybee, on the other hand, isn't particularly fond of avocado flowers and tends to be distracted if alternate nutrition sources are nearby (Ish-Am and Eiskowich 1998).

The type A and type B sex system can make the love life of an avocado tree in the crowd of the commercial grove one of romantic yearning. Your inner matchmaker cries out, "Say what? As long as there are some pollinators around, type A trees can mate with type B trees and vice versa." True enough, mating is not a problem for wild avocados in their natural populations in Mesoamerica. But you've also identified why cultivated avocado trees can be so lonely. Nowadays, in an industrial avocado planting, type A trees are challenged to find a type B mate or vice versa because such groves are planted in a single mating type for as far as the eye can see. We've got a bit of a romantic mystery here. Is that avocado baby in your hands the product of a intervarietal tryst—or it is the result of a lonely avocado tree left to its own devices?

Just like industrial bananas, avocado varieties are clonally propagated. Just like bananas, there's one favored variety. Unlike banana plants, avocado cloning involves grafting. Here's how grafting is accomplished for the avocado: Seedlings are germinated from avocado seeds (like you did when you were a kid, but not necessarily with toothpicks and in a clear glass of water). The seedling eventually becomes the root system and lower trunk for the grafted tree and is termed the rootstock. Once the seedling is established, a bud or shoot is cut from a branch of the chosen cultivar, say, an adult tree of the world's current favorite avocado, the Hass variety. The seedling's bark is sliced to accommodate the donated plant material. The donated material is affixed to the seedling with grafting tape. Properly done, with compatibility and luck, the graft and seedling grow together to join vascular systems, becoming a single plant. Once the graft has established, bursting forth with a new leafy branch, the top of the original seedling that nursed the new branch is lopped off. The process may take weeks. The graft's established branch, the scion, nursed by the rootstock, is trained to become the top of the trunk of the tree and produces all subsequent branches, leaves, flowers, and fruits. The avocado nurseryman keeps an eye on the growing tree, snipping off the occasional adventitious branch from the rootstock so that the only branches are those of the scion. The tree is like two LEGO bricks snapped together top to bottom that eventually join tissues to glue themselves to each other (Ernst, Whiley, and Bender 2013). Under the system just described, each avocado rootstock is a sexually produced individual and different from each other. But the scions are clones descended from a single mother tree.

For the commercial avocado grower, just like the commercial banana grower, the uniformity of a plantation of a single genotype has immediate benefits. All trees can be managed in the same way. Even if grown on rootstocks that are genetically different from each other, a single scion variety produces a much more uniform product than a grove of trees grown directly from sexually produced seeds, each with a different genotype. Popular woody food plants are often propagated as rootstock-scion combinations. Not only avocado varieties, but Granny Smith apples, Washington navel oranges, Cabernet Sauvignon grapes, Chickasaw pecans, and Spain's famous Picual olives are all asexually created scions that sit on rootstocks of a different genotype. For the avocado, asexual propagation started in earnest somewhat over a hundred years ago.

Clonal rootstocks can provide even more genetic-based uniformity. Certain citrus and citrus relatives regularly set apomictic seed, and these typically serve as rootstocks for citrus scions (Wutscher 1979). Clonal avocado rootstocks have become popular in certain countries in recent decades. Creating clonal avocado rootstocks involves double grafting with three participating layers: a sexually produced rootstock, a clonal interstock, and a clonal scion. Once the clonal interstock is an established graft on a seedling rootstock, it is treated so that it sprouts spontaneous roots instead of additional branches. Then the bud for the clonal scion is grafted to the interstock. The lowest layer, the original rootstock that nursed the interstock, is eventually removed in its entirety, leaving the interstock as the clonal rootstock (Ernst, Whiley, and Bender 2013). Think of stacking three LEGO bricks bottom to top. Once the top two are established, the bottom one is removed.

Clonal rootstocks can provide benefits beyond simply boosting the uniformity of an avocado grove. Often a clonal rootstock is chosen because its genotype confers resistance or tolerance to a disease organism for which the scion is vulnerable. The threat of *Phytophthora* root rot disease, a major problem for the avocado industries of more than a dozen nations, has been slowly reversed by discovering and adopting tolerant clonal rootstocks (Coffey 1987). (Want more on rootstocks and grafting? A recent accessible article by Warschefsky et al. [2016] in *Trends in Plant Science* is filled with fun facts and useful information about grafted plants.)

Let's return to the asexually propagated scions that engage in avocado sex. All scions of the same variety are the same single genotype. If the original mother tree is type A, then all of her asexually created progeny, grandchildren, and great-grandchildren are also type A. Just as in the case of the vegetatively propagated banana, all of the separate individuals of an avocado scion variety are essentially "pieces" of one single, large, dispersed genetic individual. If a grower plants a grove of trees that are a single variety that is type A, any tree within that grove will be surrounded by trees of the same mating type. All the trees in the grove flower in synchrony. A pollinator might have to fly a long way to find a tree of type B. In this situation, not only must the lonely clonally propagated avocado tree overcome the challenge of within-flower protogyny and within-tree sexual synchrony, but it must overcome the challenge of the sexual synchrony of its dispersed body (Ellstrand 1992).

Consider the type B Fuerte scion variety and the tree of choice for much of the commercial avocado industry in the mid-twentieth century. Buds taken from a seedling tree in the Mexican state of Puebla were transported to California in 1911. When grafted trees with this genotype (first named Number 15) withstood California's severe 1913 freeze, they were renamed Fuerte (in Spanish, "hardy" or "strong") (Shepherd and Bender 2002). Maybe its skin is a bit thin and requires careful handling, but Fuerte has a lot going for it. Its taste and texture are extraordinary. Fuerte is considered by some avocado aficionados to have the best flavor of all varieties. A handsome, pear-shaped, forest-green fruit goes with that great taste. Why would a grower want to grow anything else? And that's the point. Mid-twentieth-century avocado growers loved type B Fuerte—perhaps loved them too much. Many grew nothing but Fuerte. Until that time, avocado farmers experimented with a diversity of varieties, both A and B. The Fuerte was the first globally important variety commercially.

As late as 1950, the challenges of avocado sex were still not widely understood, even though New York Botanical Garden's director of laboratories Dr. Arlow Stout first recorded the curious sexuality of the avocado flower after a 1923 visit to California (Stout 1923). Thus, serious growers rapidly adopted the Fuerte without worrying about matchmaking. If a grower was an experimentalist, the nearest type A mate for her type B Fuertes might be as close as the next tree. But for growers hell-bent on getting top dollar, any tree grown that was not a Fuerte meant less than maximum income. For those ambitious Fuerte-only growers, the nearest type A tree might be, with luck, the next ranch. But as Fuerte-only groves became increasingly commonplace, and given the fact that many other avocado scion varieties are type B, just like Fuerte (see table 4.1), the nearest type A tree might be in the next township. With a single synchronous clone dominating the scene, it's kind of hard to find a mate of the alternate type. Not surprisingly, it wasn't long before some Fuerte growers began to think that they should be getting better fruit set in their groves (Hodgson 1947).

Was it a problem with the availability of suitable mates? The role of cross-pollination in the avocado fruit set was controversial among the scientists of the day. Dr. Stout (1923), the aforementioned Kinsey of the avocado, advocated interplanting type A and type B trees as complimentary pollen sources to optimize fruit set. But other experts flatly stated that intervariety pollination had nothing to do with fruit set. University of California avocado researchers Bob Bergh and Don Gustafson summarized those skeptical comments in their 1958 review:

TABLE 4.1. *Mating Types of Some Avocado Cultivars*

Type A	Type B
Green gold	Bacon
Gwen	Edranol
Hass	Ettinger
Lamb Hass	Fuerte
Lula	Nabal
MacArthur	Queen
Pinkerton	Regina
Reed	Sharwil
Topa Topa	Zutano

According to Chandler (1958), p. 213, "Fuerte seems to fruit about as well in (large solid) blocks or as single isolated trees as in closely mixed plantings. If cross-pollination increases its crops, that increase is too small to be detected among the results of other influences. . . ." Earlier, Hodgson (1930), p. 65, had written in similar vein: ". . . no case has yet been brought to light in this state where the provision of cross-pollination has measurably improved either the regularity of bearing or the amount of yield of individual trees or solid plantings of single varieties. . . ."

Interplanting or pure groves? Clearly, the results from a good experiment or experiments could help growers make a decision about how best to plant out their groves. But how to do an experiment with trees that take years to bear fruit?

Before answering this question, it is important to point out that a second timing issue confounds fruit set in avocados, potentially obscuring the effects of finding a mate. A natural process that occurs in even some wild trees, alternate bearing is the bane of fruit and nut tree farmers: from apples to apricots, from pecans to pistachios, and from olives to mandarins, alternate bearing is a continuing cycle of a season of heavy yield (an on year of abundant fruit set) that seesaws biennially with a very light season (an off year of almost no fruit set). The cycle starts when a tree is stimulated to produce extremely low or high numbers of flowers or fruits. Once started, alternate bearing becomes fixed until the next jolt comes along and resets a new alternate-bearing cycle.

Most avocado varieties are prone to alternate bearing. Imagine a cohort of young avocado trees that reach reproductive age simultaneously. Their year-to-year fruit set may be more or less stable until the trees are hit by, for example, a freeze that doesn't kill the trees but stresses them to the extent that they drop most of their fruits, creating an off year. The following year is on; the next, off, and so on. Alternate bearing occurs because there is a physiological-based negative relationship between the number of hanging fruit and the number of reproductive buds that develop for the next year. An off year created by a stress allows a reproductive bud production boom—and the cycle is set (Salazar-García, Garner, and Lovatt 2013).

One month after California's severe freeze of 2007, I drove US Highway 101 past mile after mile of blackened avocado trees from inland Santa Barbara County through San Luis Obispo County. Roughly half of that year's crop was lost in the region. President George W. Bush declared the agricultural losses a major disaster (Carman and Sexton 2007).

When the damage is regional, all the groves get reset to the same alternate-bearing cycle. Thus, farmers of the entire region not only suffer from the freeze year, but they also share the pain into future years. All their groves start alternating in tandem. The economics of regional alternate bearing are grim. Avocado orchards in a region that suffered from the same freeze end up synchronized, sharing very low-yield off years when avocados are rare and prices are high for the few fruits produced and alternating with years of excellent yield (on years) but miserable prices because everyone has a bumper crop. When a large region suffers from alternate bearing, nobody in the supply chain benefits, except customers in an on year.

Alternate bearing is unusually problematic for Fuerte. Not only is it sensitive to environmental stresses, but even the first good year of production is sufficient to start alternate bearing in this variety (Hodgson 1947). Alternate bearing continues to be a problem for avocado growers today, but scientists are beginning to offer managers some strategies for manipulating crop load, from adjusting timing of harvest to applying plant-growth regulators (originally called plant hormones) (Whiley, Wolstenholme, and Bender 2013).

Strong alternate bearing frustrates attempts to measure the influence of other factors on fruit set. In some off years, yields are so close to zero that measuring the influence of intervarietal pollination on fruit set would be impossible. Other factors can blur the relationship, such as variability in fruit set due to pollinator availability, insufficient watering, or effects of different rootstock genotypes (Whiley, Wolstenholme, and Bender 2013). No wonder so few avocado scientists had the courage to collect data to try to disentangle the effects of A-B pollination from the other causes that can contribute to variation in avocado fruit set.

Nonetheless, the problem is testable. If intervarietal pollination plays an important role in boosting fruit set, then trees that cross with another variety should have a higher fruit set compared to ones that do not cross. Ideally, if one could infer relative paternal contribution, then the relationship between avocado crossing and fruit set could be tested. Avocado researchers of the mid-twentieth century used ingenious methods to get at the question. For example, Bob Bergh and his colleagues in California studied groves where Fuertes grew at varying distances from type A varieties (in one case, for example, the type A variety Topa Topa was grown as a windbreak). They reasoned that a shorter distance to the alternate-mating type increased chances for intervarietal romance. They found that the Fuertes closest to the type A varieties typically had higher yields (averaging a 40% yield boost) than those more distant, often with yields that decreased with distance. In another experiment, they manipulated their trees to make the crossing distance shorter still. They grafted branches of an array of different type A varieties onto different Fuerte trees. The fruit set on the grafted trees was recorded for six years and compared to an ungrafted control. In the majority of cases, fruit set on the grafted Fuertes was dramatically greater than the corresponding ungrafted tree (Bergh and Gustafson 1966). Bergh (1968) boldly summarized these results in his first paragraph reviewing the data (I will hold off quoting from his second paragraph for a moment because that passage becomes especially meaningful by the end of the chapter):

> It MUST now be regarded as an established fact that avocado trees set more fruit when there are flowers of a different avocado variety nearby. How do we know? From actual counts of the numbers of fruits on the trees compared.

End of story?

Collectively, these data sets argued strongly for a relationship between outcrossing and fruit set in the avocado. But they did not directly measure whether the fruits on the grafted trees or those adjacent to an alternate variety were actually the result of intervarietal crossing. It is always possible that something else was driving the observed differences. For example, could grafting stimulate fruit set? To obtain a clearer answer, you would need to be able to identify the father of a fruit.

In high school Latin class, I was taught the phrase "*Identitatus patris incertus semper est*," that is, "The identity of the father is always uncertain." To the best of my memory, I was taught that phrase by a fellow student, not the teacher. It makes sense because that was not the sort of idea to be shared in the classrooms of the 1960s! So let's attribute that one to the student (now a professor of botany; who says botanists don't have a sense of humor?). Whether the Romans used that phrase or not, the point is clear. You can pretty much be sure who the mother of a child is, especially at the time of the birth. By then, the biological father may be long gone. The phrase is even more accurate for the seeds of flowering plants. The seed inside the fruit hanging on the tree is attached to its mother and surrounded by maternal tissue. To identify the father, you would need to do genetic analysis to compare the mother plant, the baby, and the possible fathers. With enough markers available for enough genes, a geneticist can perform paternity analysis by comparing the genotype of the seed with those of its possible parents, thereby identifying seeds whose male parent differed from the mother's variety. Genetically based markers eventually became available in the late 1970s that could be assayed in both adult tissues and seeds.

The procedure is straightforward. First, multigene profiles of a seed and its mother are obtained. The avocado mother is also a possible father. So the geneticist also screens all other possible male parental varieties. The scientist subtracts the mother's genetic contribution to the seed. The contribution that's left came from the father. Comparing that genotype to the potential fathers, some potential paternal varieties are eliminated. If enough information is available, only the real father remains on the list. The approach is essentially the same as paternity tests in humans to resolve the identity of a child's father. In the case of an avocado, such genetic analysis can additionally identify when the mother (or one of the mother's identical sisters in the grove) is also the father (note that the avocado can engage in three kinds of effective selfing: autogamy, geitonogamy, and within-variety inter-tree crossing).

The first avocado paternity tests were conducted by Michelle Vrecenar-Gadus, a graduate student at the University of California, Riverside (UCR). Like UCR's Bob Bergh before her, she sought out a natural experiment to conduct her work. She found two similar groves that were on the same California avocado ranch: one grove was planted entirely in the Hass variety; in the other, each Hass (type A) row alternated with a row of Bacon (type B) trees. She sampled Hass trees in each grove to estimate the number of mature fruit produced. Then she harvested ten fruits from each tree, genetically analyzed the seeds in the fruits, and compared the seeds' genotypes to the Hass mother and to the potential Bacon father. Not surprisingly, the yield in the mixed grove was substantially higher than that in the pure grove. Almost 90% of the Hass fruits in the mixed grove were fathered by a

Bacon tree, and the rest were fathered by a Hass tree. Over all the trees sampled, yield increased with intervarietal crossing rate. But tree-to-tree yield varied tremendously such that the outcrossing rate explained only about 10% of the yield variation. Surprisingly, the pure Hass grove produced a substantial fraction of outcrossed fruits, about 42%. That's a lot of fruits given that the nearest Bacon tree to the grove was over 260 feet away. Overall, outcrossing played a significant role in the paternal parenthood of fruit in both the mixed and the pure groves (Vrecenar-Gadus and Ellstrand 1985).

More than a half-dozen genetic studies followed by numerous scientists in avocado-growing countries. Most studies found a positive relationship between outcrossing and yield. But the results varied considerably. For some, the relationship was statistically and economically significant (e.g., Degani, El-Batsri, and Gazit 1997); in others, it was not (e.g., Garner et al. 2008). The largest and most comprehensive study was published by a UCR team led by Marilyn Kobayashi in the lab of Michael Clegg (Kobayashi et al. 2000). This tour de force involved the genetically analysis of almost 2,400 Hass seeds collected from seven groves over four years in two different climactic regions of California. The team used a lot of genetic markers relative to prior studies, not only enough to distinguish between selfed and outcrossed seeds, but enough to distinguish between four possible paternal parents: selfed from Hass, and natural crosses with Fuerte, Bacon, and Zutano. Overall, they found a significant correlation of a tree's outcrossing rate to its fruit production. Closer inspection of the data showed that the trend was driven by three of the four coastal groves that showed the effect strongly (accounting for about 25% of the variation in yield). For the other four groves, there was a positive, but not statistically significant, relationship between outcrossing and yield.

Contrary to what you learned in school, science is not always tidy. The house built by research is always in need of some work. The mystery of avocado sex and yield is one of many in science in which no popular consensus has emerged. A half century earlier, Bob Bergh nailed the untidiness in science as it relates to the mystery of the avocado's romances. Now read the *second* paragraph of his 1968 review on outcrossing and avocado fruit set:

> This may not be true for all varieties. And it may not be true in all avocado regions. And it may not be true for a given variety in all years—in fact we have found great variations in the cross-pollination effect from year to year.

Before drawing our final conclusions about what all of these scientific data mean regarding sex and the single avocado, we need to address a question that you are probably asking: Why are these recent studies involving genetic markers focusing on Hass, rather than Fuerte? By the 1980s, Hass production eclipsed Fuerte in California. Today roughly 90% of the world's industrial avocado production is based on Hass (Schaffer, Wolstenholme, and Whiley 2013). What happened to Fuerte? The reason is, in part, due to the avocado's third and final reproductive timing issue. Specifically, when is an avocado ripe? Answering that question is not easy. For some species, if a fruit looks good enough to eat, it actually is; whereas in other species, such as the avocado, beauty may be only skin deep.

Consider the following story, but do not try this at home. A young plant scientist fought early morning traffic to make his way to a university agricultural field station. He had skipped breakfast to get on the road early to miss the worst of rush hour. He jumped out of the car to a stunning variety of collections of subtropical fruit trees—dozens of varieties of avocados, cherimoyas, macadamias, guavas, kiwis, and persimmons.

Time to get to work collecting data in the cherimoya collection.

But he was hungry. Increasingly faint from half a day without food, he approached the persimmon trees under the noonday sun. His hungry stomach constricted with a spasm. His thoughts turned to the plump fruits at the edge of the road:

Why not grab a persimmon off the tree for a snack? Wait, don't persimmon varieties need to ripen after picking? Some do. But others ready to eat off the tree. Bought a Fuyu persimmon at Safeway last week; that one doesn't need to ripen off the tree. What did it look like? Squarish, flattened-pumpkin-shaped. Here's one that looks like a Fuyu, I think. Pumpkin orange and sort of pumpkin-shaped. Right size, too. Sure? Hard to say; I don't have a map for this collection. Ow. Another stomach spasm. That Fuyu the other day was as sweet and crunchy as an apple. Why not a nibble?

His head snapped back. An intense sensation signaled from parts of lips, tongue, and palate blasted by protein-denaturing tannins. Extreme astringency. Not exactly pain, but immediate numbing and causing a dry pucker similar to that elicited by strong black tea or too young Cabernet Sauvignon, but intensified the way that a pat on a cheek might be compared to a blow by a baseball bat. After a few minutes, the five-alarm feeling was subsiding without lasting effects.

A little lesson. *Ah, there are varieties that look like Fuyu but act like Hachiya!*

He never again made that mistake.

I was that young scientist.

Just because a fruit looks palatable, doesn't mean that it is. The well-known Hachiya persimmon must first be picked and then allowed to ripen to a near-soupy consistency to enjoy it without a trace of astringency. Yes, Fuyu persimmons can be eaten off the tree, but other varieties that look just like Fuyu—some of which were in that collection—must be picked when mature and then properly ripened, just like Hachiya.

Mature? What is the relationship between maturity and ripeness? We all know that if we buy a grape or orange it will be ready to eat. That's because these fruits come from species in which maturity and ripeness occur simultaneously. For that kind of fruit, apple, for instance, at the right point in time, you can pick the fruit off the plant and eat it. If you pick an apple or orange or grape too early—that is, unripe—it often doesn't look right, and it certainly doesn't taste good.

We know that bananas are different. Usually, when you bring the fruit home, it is not ripe. Wait a few days; the banana ripens and becomes edible. We have learned in the last few decades that an unripe, hard tomato should be treated in the same way as an unripe banana. The technical term for species whose unripe fruits can ripen off the maternal plant—like the banana and the tomato—is climacteric (think "climax"). These contrast with the pick-and-eat non-climacteric apples, grapes, and oranges. Well-known climacteric fruits include apricots, kiwis, nectarines, and peaches. Sometimes it varies within a species. As we well know, some persimmon varieties are climacteric (Hachiya);

others are not (Fuyu). A climacteric fruit must reach maturity before being harvested, or it will never ripen properly off the tree. An immature fruit, one that is picked too early, may look tasty but is asking for trouble. Some fruits can be picked immature and treated to undergo a kind of pseudo-ripening. With the right treatment, some do very well. But some can be disasters. You've undoubtedly had a bright red slice of tomato on a burger in January that had less taste than the burger's polystyrene foam clamshell, but the same texture. Bad things happen for those who sit and wait for immature pears, apricots, and, yes, avocados to ripen.

Avocado is a climacteric species. Picked immature, the bad news varies with variety. Avocado aficionado Mrs. B. H. Sharples may have said it best in 1919:

> Nature has chosen to clothe this choice gift to man in sombre garb, and the public buys the avocado, not because of its appeal to the eye, but on the recommendation of a friend, or because he has experienced for him-self the pleasure and satisfaction of eating the ripe avocado at its best.
>
> The immature avocado has not the delicate blush of the half ripe strawberry, to catch the eye of the purchaser, or the alluring "sweated gold" of the green orange, nor the flaming invitation of the unripe persimmon.
>
> Gullible man is enticed again and again to buy these acid, puckering, disappointments, because of his inherent conviction that beauty cannot be false, but one flat, insipid avocado that has been rushed into the market prematurely will make him wary of the most tempting display her worthy sisters can make, in their modest gowns.
>
> With some varieties of the avocado, the immature fruit mellows evenly after taken from the tree and reaches the

> public in a very nice condition in so far as appearance is concerned; the flat, watery flavor or "cucumber flavor" being the only evidence that it was picked too soon.
>
> With other varieties the skin assumes a withered, wrinkled appearance, while the flesh mellows evenly, as in the fully ripe fruit. Others never become mellow when taken from the tree too soon but, after a few days become leathery, tough and inedible.

At best a watery cucumber, at worst an inedible chunk of hard rubber—immature avocados are unforgettable. And Mrs. Sharples got it right; it is avocados "rushed into the market prematurely" that can poison the marketplace.

What do premature harvest and immaturity have to do with the rise of Hass and the fall of Fuerte? By the 1960s, avocado growers and packers looking for a global market needed a variety with a skin thick enough to survive long-distance, perhaps intercontinental, shipping. To avoid competition with the abundant and popular Fuerte, they were also looking for varieties that could come to market before or after Fuerte season. To some, the Hass variety seemed ideal. Its fruits have a thick but easy peeling skin. Hass fruit production overlaps with Fuerte but also goes considerably later than Fuerte. Growers can "store" mature Hass fruits by leaving them on the tree. Hass fruits are available for harvest long after the year's final Fuerte fruits have dropped to the ground and have been savored by the avocado rancher's dogs. Bob Bergh (1968) went so far as to suggest that type A Hass could be interplanted with type B Fuerte to romance the Fuerte into higher yield. But to others Hass had an unforgivable flaw, a flaw so deep that only a father's love for his children saved the original Hass tree from the ax.

Like Fuerte before it, Hass started as a seedling tree. When California mail carrier Rudolph Hass bought three seedling trees in 1926 to serve as rootstocks for Fuerte, he could not have guessed that one would serve as the future of the world's avocado industry. In fact, it was the only of the three that refused to accept the Fuerte graft. Hass let the seedling grow but neglected it. In fact, Mr. Hass considered the fruit to be visually repulsive. In contrast with the popular Fuerte's smooth, green, pear-shaped orbs, the seedling's mature pebbly-skinned fruits were not only dusted with black, but when ripe, they turned purple-black. When his children begged him ("You gotta try it, Dad!") into giving the grenade-shaped fruit a chance, the creamy, nutty flavor (18% oil) changed his mind. So did its extraordinarily long season. Naming the selection after himself, he received a US Plant Patent in 1935 and attempted commercialization (Shepherd and Bender 2002).

The Ugly Duckling fruit that first repelled Mr. Hass had the same off-putting effect on consumers. Visually, the Hass was everything the Fuerte was not. In fact, consumers well knew that a black-skinned Fuerte is a rotten Fuerte. All of the major varieties of the time were also green-skinned until they rotted into blackness. Hass stood out like a black (and presumed rotten) sheep. You can imagine that Bergh's suggestion to interplant Fuerte with Hass was met by some with amusement, if not derision.

While Hass was meeting opposition as a post-Fuerte choice, the avocado industry sought a variety to fill the pre-Fuerte market. In particular, California growers were motivated to get a fruit to kick off the avocado season. The season's first fruits in the market command the highest prices because competition is so low and the demand so high. It must have rankled to see the fluorescent-green Lula from Florida as an avocado choice in California groceries in the fall. Green-skins like Bacon, Zutano, and Pinkerton were considered as pre-Fuerte candidates for late fall and early winter. Consequently, some unscrupulous California growers placed immature green-skins on the market early in California and particularly in the Midwest, where consumers were relatively avocado naive. A green-skinned Bacon or Zutano fruit, when picked mature and properly ripened, is a good avocado. But immature Bacons picked and marketed in September and October of the latter years of the 1970s and early years of the 1980s were ghastly, partially "ripening" to semi-edible, non-palatable, uneven rubberiness. Customers, even in Illinois, were not pleased.

In those days, fall was more or less an avocado desert. After months without avocados, the first fruits of the new season set consumer perceptions. Enough immature Bacon and Zutano fruit pushed into the market rapidly educated shoppers to stay away from green-skinned fruit. Months later, when Fuerte finally arrived in the produce section, they were shunned as well. The immature green-skins poisoned the market and inadvertently deposed the Fuerte as the king of avocados. After all, it takes a trained eye to tell apart the green-skin varieties.

But when Hass appeared later yet, it looked different enough for consumers to give it a second chance; after all, many remembered that avocados *used* to taste good. Year by year, the shift away from green-skins to alligator-skinned, dusky Hass progressed. Green-skins didn't stand a chance. Hass can be harvested early enough to pick up a good chunk of the Fuerte season. If consumers were going to avoid Fuertes, then Fuerte growers decided that they might as well convert to Hass. As Hass became better known, the green-skins—Fuerte, Bacon, Zutano, Pinkerton, and more—suffered. In fact, because of the ability of Hass to be saved on the tree until harvest, it could be held through the summer until fall. Hass became the post-Fuerte choice as well as the pre-Fuerte choice. Perhaps jet-black, worn, and deflated, the last of the over-mature Hass fruit are a bit weary but are still relatively tasty (accumulating a very high oil content over time on the tree) compared to the immature green-skin autumn alternatives. Decent mature Hass avocados can be produced by a tree for much more than half a year. In addition, when local Hass are out of season, fruits of the variety can be shipped in from an alternate location with a different season (for the United States, it is often Mexico).

The global avocado industry is dominated by one variety. The domination is far from as absolute as what the Cavendish has done to the global banana. Nonetheless, the world's avocado behemoth, Mexico, produces primarily Hass (Flores 2015); so do New Zealand and Kenya. In California, 95% of avocados grown are Hass. The fruit is the favorite of North American and western European consumers.[2] Most growers, packers, and grocers have welcomed the change to one genotype. After all, as we learned with the banana, it is easier to grow, package, transport, and market uniformity than diversity.

There remain regions that are havens for other varieties. Eastern Europe still likes green-skins. Argentines have a love affair with the massive Torres (almost two pounds of guacamole from a single fruit). For American foodies who crave

2 See "Overview Global Avocado Market," www.freshplaza.com/article/156557/OVERVIEW-GLOBAL-AVOCADO-MARKET; and "Hass Mother Tree," www.avocadocentral.com/about-hass-avocados/hass-mother-tree.

diversity, it is still possible to find varieties beyond Hass. In late summer, some stores and certain farmers' markets offer a green-skin roughly the size and shape of a softball. The variety is called Reed. (I never met a Reed I didn't like.) Zutanos, Bacons, Fuertes, and others still persist in the United States and elsewhere. The second most important avocado state, Florida, produces the shiny green-skinned West Indian subspecies. This group, such as the provocative Lula, is well-suited for semi-tropical and tropical climates. The other avocado-producing state of note, Hawai'i, lags far behind the others, featuring an exotic array of locally adapted varieties that end up savored by the natives and the abundant tourists. (State rankings of avocado and other American products can be gleaned from the National Agricultural Statistics Service website, www.nass.usda.gov, the US equivalent of the global FAOSTAT website that I fawned over in chapter 3.)

To sum up, for the global market, Hass started as a mid-season, easily transported fruit that raised noses at first and eventually eyebrows. In North America, as green-skins fell into disfavor, early but adequately mature Hass displaced Fuerte over much of its season. At the same time, consumers found that very mature late California Hass fruits and those shipped from elsewhere were highly superior to extremely early, immature, foul-tasting, semi-crunchy Bacons. Thus, by the time that genetic markers were ready for outcrossing studies, interest in the relationship between outcrossing and Fuerte yield was gone.

Time to return to our romantic mystery.

Avocado fruit set varies tremendously between trees and between groves. Most avocado scientists admit that outcrossing influences yield in a small but positive way (about

10%). Studies have demonstrated that the strongest relationship between outcrossing and yield occurs for the maternal tree that is immediately adjacent to one of a different mating type. To get a substantial per tree yield boost, it is probably necessary to plant orchards so that every other tree serves as a "pollinizer," that is, a pollen source (Salazar-García, Garner, and Lovatt 2013).

Growers are reluctant to replace a Hass tree with a non-Hass pollinizing variety. Hass fruits command a superior price in the global market. Nowadays, the view is that the trade-off of increased Hass fruit production in an interplanted grove is insufficient to justify replacing a lot of Hass trees with green-skin type B pollinizers that produce hard-to-market fruits (Salazar-García, Garner, and Lovatt 2013). From the point of view of industrial growers, it's not worth it to interplant. But what about the point of view of the tree? Are self-pollen and cross-pollen pretty much six of one and half a dozen of the other? In the case of the avocado, is Darwin's dictum wrong? Clearly, avocado trees go to extraordinary lengths to mate with someone genotypically different from themselves. The timing of sex expression on individual flowers discourages within-flower selfing. The synchrony of whole-tree sex expression discourages selfing within a tree—and between genotypically identical trees of the same cultivar. Is it all for naught in the industrial Hass grove?

Have we completed our detective work, or have we missed something about the romantic biology of the avocado? Thus far, we've only discussed the paternity of mature fruits on the tree. That approach assumes the proportion of paternal genotypes doesn't change after fertilization. That is, the initial outcrossing rate is the same as the outcrossing rate measured on mature fruits. But, as we know, for every mature fruit, about a thousand fertilized fruits have already dropped. Are paternity patterns for harvested fruit different from the paternity patterns at fertilization?

Shmuel Gazit, Chemda Degani, and colleagues, avocado researchers at Israel's Volcani Center, used genetic markers to compare the paternities of aborted (botanically, "abscised") fruitlets with those that make it to harvest (Degani et al. 1986; Degani, Goldring, and Gazit 1989; Degani, El-Batsri, and Gazit 1997). Paternities of the seeds in dropped fruitlets are different from those of mature fruits. They found that outcrossed fruits are preferentially held on the tree. In one study, less than one-quarter of the fruitlets that dropped after a month of development were the result of outcrossing to a different variety; but for the mature fruits, 84% were the result of intervarietal crossing. Thus, avocado trees not only have floral timing mechanisms for avoiding self-fertilization; once fertilization has occurred, they boost the fraction of outcrossed embryos at the expense of the lives of thousands of their selfed siblings. The Israeli scientists reported that within-genotype fertilizations occurred at surprisingly high rates, but the trees typically preferentially shed selfed embryos; this radically increased the fraction of intergenotype outcrosses in mature seeds, leading to high rates of apparent outcrossing among the mature fruits. Avocado trees indeed go to great lengths to have a lot of outcrossed progeny. The trees do care. When given a choice, they do their darnedest to put their energy into real outcrosses. (OK, I know that trees don't "care," but you get the point.)

Planting hundreds of thousands of trees of the same variety gives the trees little choice, just as the typical urban avocado consumer has a decreasing choice of avocado varieties in large commercial markets. If the avocado story is hauntingly similar to that of the banana, it is because of the global trend that favors the economy of genotypic uniformity. We have lost easy access to many excellent avocado varieties because of the "one-size-fits-all" attitude of those who influence the market. But make no mistake: a properly picked, properly ripened, testicle-reminiscent Hass is a *great* piece of fruit.

What can we take home from our scientific voyeuristic adventure regarding the romance of the avocado? We started with *When Harry Met Sally* and, yes, we found how important timing is for avocado sex and fruitfulness. But our conclusion may be likened to yet another romantic comedy, Nancy Meyers's 2009 film *It's Complicated*, because the biology of romance is, much like romance itself, complicated.

The next chapter moves on to a crop whose sex life is anything but lonely, a wind-pollinated plant that can mate with others that are as distant as a half mile or more. Plant breeders have worked with the sugar beet's promiscuity to build a better sugar beet. But sugar beets have behaved badly, sexually engaging with not-so-nice suitors, and sugar beet farmers have paid the price.

Avocado Toast—It's Not Just for Breakfast Anymore

Leslie Leavens should know what to do with an avocado. The Leavens Ranches that she manages with her family has grown them, as well as lemons and more, for decades. In her "spare time," tireless Leslie wears many hats, including working to create good and affordable housing for farmworkers in California's Ventura County. Past president of the

Farm Bureau of Ventura County, Leslie will have taken on more side projects by the time you are reading this (Warring 2013).

She taught the following recipe to my wife during a mini-reunion getaway with their California Agricultural Leadership Program buddies.

When you prepare your avocados for this recipe, remember the role of good timing that got you the fruit in your hand.

2 or 3 fully ripe, but not overripe avocados, peeled and sliced
Slices of your favorite bread for toasting
One or two halved Mexican (aka key) limes or quartered Persian (aka Tahitian) limes
Good, fresh chili powder (or your favorite spicy mixed seasoning)
Optional: Bacon slices, tomato slices, sautéed mushrooms, etc.

Number of servings depends on the size of your avocados, bread, hunger, and so on.

Toast the bread. Spread avocado thickly on the still-warm toast. Squeeze lime juice, and sprinkle chili powder to taste. Move quickly so that the toast is still warm as you enjoy the aromatic combination of flavors.

5 *Beet*

PHILANDER (FEMALE) AND PHILANDERER (MALE)

Roses are red. Violets are blue. Sugar is sweet and comes from a beet.

When it comes to romance, sweetness usually carries the day. Nothing says romance like, for instance, a box of chocolate truffles. But it wasn't always this easy. In the natural world, sweet foods are hard to come by; candy-sweet foods are almost unknown. As recently as a few millennia ago, much of the civilized world was having a hard time finding sweet-tasting edibles. Dried fruits were probably the most common of that scarce community. Fresh fruits are seasonal and have a short shelf life. Until honeybee domestication, honey harvesting was at best challenging and difficult work and, at worst, dangerous. Even once honeybees were domesticated, apiculture was restricted to temperate Eurasia and northern Africa. That's about it. Is it any wonder that the Aztecs served their chocolate as a bitter concoction of various ground seeds, vanilla, chili, and other spices? But at the other end of the globe, a sweet revolution was slowly spreading.

Just about any kid growing up where grass abounds—meadows, lawns, prairies, savannahs—will at one time pluck a stem and chew on it. It's not hard to imagine a bored and hungry Neolithic teen hiking home under the hot sun after a long morning of foraging roots and nuts for her family. She uses her obsidian knife, a gift from her uncle, to cut a chunk of succulent stem from a tall grass for a chaw. Mashing the fibrous tissue with her molars, her mouth is flooded with sweetness. *Well, now, THAT's a surprise!* After enjoying a session of juicy mastication, she cuts several hand-long stems pieces to share with her siblings and friends in the village. Otherwise, would they ever believe her that something could taste *so* good? (Not nearly as good as the modern domesticated version, but good enough.) *Wait, two more stem chunks to share with that cute farmer with the healthy pigs!*

The first "sweetie"?

That's an approximation of the event on the island now known as New Guinea that changed, and is still changing, history. For thousands of years, only the people of that island and, eventually, the Far East and Oceania enjoyed the products of the crop domesticated from that species and some close relatives. The term sugarcane is now applied to any of the domesticated perennial grasses of the humid tropics in the genus *Saccharum* (as well as their interspecies hybrids) (Roach 1995). First cane, then cane juice, and finally a more easily stored and transported product. Roughly three thousand years ago, some South Asians might have been the first to thoroughly dry cane juice and be pleased with the remaining brown crystals. The contemporary highly purified industrial white version is what we call table sugar, the chemical compound sucrose (Bakker 1999).

Sugarcane diffused oh-so-slowly west, making it to Persia around 1,500 years ago, creeping into Europe a few hundred years later, during the Dark Ages. This equatorial plant requires at least a semi-tropical, and preferably tropical, climate. Consequently, cultivated sugarcane held on by its fingernails in the warmest, most protected parts of the Mediterranean basin. The European colonization of sub-Saharan Africa and the subsequent discovery of the New World tropics offered European entrepreneurs opportunities for cultivation on a grand commercial scale. Columbus himself introduced the big grass into the Caribbean as a plantation crop. Dry brown unrefined sugar loaves were transported back to Europe. By then, the product was pure enough that spoilage was difficult. Soon, a crude version of what we know now as sugar was produced in sufficient volume to become the world's staple sweetener (Bakker 1999; Blackburn 1984).

By the 1700s sugarcane plantations were a keystone of pan-tropical imperialism, particularly so for Portugal, Spain, France, and Britain. They were also a component of Britain's notorious "triangle trade." A quick refresher from high school American history: The first leg of the triangle was the transport of West African slaves to the Americas. The slaves who survived the horrific journey fueled the New World plantations with their skill and muscle. The second leg involved shipping raw plantation products—not only unrefined sugar, but products like tobacco and cotton—to the British Isles. There, factories added value to those raw materials, creating not only refined sugar, but rum (from molasses, a by-product of sugar refining), textiles, and other manufactured goods to be sent on the third leg to Africa to be traded for more slaves (Smith 2013).

Although rum was a significant by-product of sugar production (for more on rum, see chapter 6 of Standage's 2005 *A History of the World in 6 Glasses*), the real thing (or its sweet molasses by-product) was even more popular among

Europeans and their well-off descendants in the colonies. Whether in a caffeinated beverage of choice, baked goods, or even meat dishes, upper-class folks had found their sweet tooth. The classic suggestion "Let them eat cake" could have only been made after cane sugar was readily available for the creation of pastries.

Sucrose as sugar became the first global industrial biochemical. (That statement could be argued, depending on whether you count distilled spirits, largely a mixture of water plus an enhanced percentage of ethanol via the distillation process. But the purity of ethanol in typical spirits of the time was far from the purity of sucrose in refined early Industrial Revolution sugar.) Technology continued to improve so that sugar from factories was increasingly pure, which also means increasingly white and increasingly resistant to spoiling. Pure sugar—that is, highly refined sucrose—is almost as sterile as pure salt. In itself, pure sucrose is nutritionally insufficient to be a foodstuff for most microorganisms. Thus, crystalline sugar, if kept dry, is easy to store and ship. People who are serious about canning well know that sugar acts as a preservative. (My favorite: fresh salmon cured by a dry pack of sugar, salt, and dill to create the Scandinavian delicacy *gravlax*.)

Speaking of sterility, cultivated sugarcane's sex life is a dull one. The wild ancestors of sugarcane are, like all grasses, wind-pollinated. These perennials are capable of both sex and some vegetative reproduction. But for the various domesticated varieties, sex largely has come to an end just as sure as if growers had chastity-belted the plants. The crop is harvested annually, before it has a chance to flower. The crop is propagated vegetatively by stem cuttings. (Clones again!) The most important contemporary sugarcane varieties are interspecies hybrids, typically as sexually active as a brick and almost as sterile. Domesticated *Saccharum* sex is largely the domain of breeders who maintain and use sexually fertile lines to create new varieties (Bakker 1999).

Considerably more sexually active is the sugar beet, now sugarcane's prime competitor. The story of the rise of the sugar beet illustrates how humans managed a plant's sex to create a new crop, in this case, to create a new one out of an old one. Recently, the sugar beet behaving badly sexually has played a role in the evolution of what is probably the worst weed in European agricultural history, demonstrating how unintended sex can create an agronomic nightmare. The two stories are intertwined. In fact, both stories share the same prologue. . . .

The sugar beet was born from a cascade of geopolitical incidents. The first tipped domino was the Battle of Trafalgar of 1805. When Lord Nelson defeated the combined Atlantic fleets of Spain and the Napoleonic empire, the victorious Royal British Navy was free to blockade the coasts of France. In retaliation, Napoleon the First issued his Berlin Decree, banning the import of British goods into any European nation allied with France, essentially the entire continent. In turn, the Royal British Navy was assigned to disrupt the shipping routes between France and her extra-continental empire. Thus, the French empire and its European allies were denied access to cane sugar from the tropics, the most important colonial-based commodity. The empire's stores of sugar were rapidly depleted. Sugar became a luxury item as its price skyrocketed. Citizens were outraged (Francis 2006). Napoleon was in no position to tell them to "eat cake." What to do?

The French government hurried to create a scientific commission to find a replacement for cane sugar. A few of those scientists knew about the previous research of German scientists to find an alternative source of sugar. Roughly a half century earlier, the German chemist Andreas Marggraf isolated sugar crystals from beetroot juice. Under his microscope, they appeared identical to sugar crystals from cane. The yield was low, only about 1.6% of the root's fresh weight (Francis 2006). His student Franz Achard carried on the work, screening all kinds of beetroots until he discovered a white one grown for animal fodder that had superior sugar content. With a grant from the Prussian government, he built a prototype to process beets for sugar. Despite a 4% yield from his beet, Achard was disappointed. Nonetheless, he had proof of concept that sugar could be extracted from a temperate European crop (Francis 2006). Achard's results were published three years before the British-Napoleonic reciprocal blockades. Interestingly, Achard claimed that he was subsequently approached by representatives of cane sugar refineries who asked him to retract his findings. He refused. A few years later his factory burned down (Francis 2006). A coincidence?

The French commission successfully repeated Achard's work. In January 1811, a French-created loaf of beet sugar was presented to the emperor himself. Napoleon was thrilled. He decreed beets be grown for sugar in France and elsewhere in his European empire. He didn't stop there. The decree included that scientific schools be created for the study and improvement of sugar-producing beets. One year later came the fall of the final domino: Napoleon's second sugar decree, calling for even more acreage to be planted with beets as well as the creation of dozens of beet sugar factories. The beet as a sugar crop was on its way (Francis 2006).

Superficially, these turn of events may not be intuitive. The beet seems an unlikely choice. Visually, compared to the colorful beets prepared Harvard-style on your plate, the clay-colored and grotesque-shaped sugar beet is a bit of a sow's ear pulled out of a silk purse. Waiting for processing, the mountains of tan elongate roots call to mind hills of dirt clods. Closer inspection of the off-white flesh of this foot-long produce evokes an exotic starchy tropical staple like *igname* or manioc. The homely sugar beet just doesn't *look* like the stuff that should be at the core of a cherry-red heart-shaped box of Godiva for Valentine's Day, Easter jelly beans, Fourth of July cotton candy, Halloween treats, chocolate Hanukkah gelt, and Christmas candy canes. But beet sugar is listed right there on the package of the divine Ritter Sport dark chocolate with whole hazelnuts bar that I finished off contemporaneously with this paragraph. Where did this plant come from? Is there any sense to why the beet became a sugar source?

The beet, *Beta vulgaris* ssp. (subspecies) *vulgaris*, was originally domesticated from the populations of a wild plant whose green leaves were harvested to be used as a cooked vegetable just like the beet's relative spinach. Such leafy greens for cooking are known as potherbs. The wild progenitor, known as the sea beet (*Beta vulgaris* ssp. *maritima*), is a scruffy inhabitant of Europe's Atlantic and North Sea shores as well as those of the greater Mediterranean. The sea beet doesn't look much like other species where it grows, and, until the roots get woody, the whole plant is entirely edible.

The cooked leaves taste a lot like spinach. Easy to see why it is a contemporary forager's favorite—and was so several thousand years ago (Biancardi, Panella, and Lewellen 2012).

Like the approximately dozen other *Beta* species, the sea beet and its domesticated descendants belong to the Amaranthaceae, a largish plant family that holds roughly two thousand described species. Amaranthaceae is a bag of mostly herbaceous plants. Most are insect-pollinated. But some, including beets, are primarily wind-pollinated. And as we know from grasses, such flowers are simple and small. The beet's bisexual flowers are the smallest yet featured in this book, only about 5 millimeters (0.2 inches) in diameter (Biancardi et al. 2005). Their flowers are typical numerically for the Amaranthaceae, with five leaf-green tepals. There's apparent controversy about whether the Amaranthaceous perianth is composed of sepals with petals absent or petals with sepals absent. *Oh, you botanists!* Let's call them "tepals" (the trend in recent scholarly papers) and move on. The three fused carpels develop into a dry single-seed fruit.

The family is important to humankind in a lot of ways and includes some notable ornamentals. Although the individual flowers are inconspicuous, they are sometimes colorful and arranged into conspicuous inflorescences, for example, ornamental celosias, as bright as crayons and as fuzzy as teddy bears. Amaranthaceae is home to lots of notorious weeds, such as tumbleweed. The final is overrepresented on the "World's Worst Weed List" (Holm et al. 1977). In addition to beets, Amarathaceous crops include high-protein pseudo-cereals (e.g., quinoa and grain amaranth) and greens (e.g., spinach and leaf amaranth). Some species are versatile: *Celosia argentea* is both an ornamental and an important African leaf vegetable. Also, some of the weeds—like young

FIGURE 5.1 Beet inflorescences. Detailed frontal view of a beet flower, looking directly into the flower reveals the perianth of five tepals. Within the tepals is the androecium of five stamens, surrounding the stigmas of tricarpellate pistil. Background, floral spikes of a *Beta vulgaris* plant.

lamb's quarters—will do nicely as an addition to a salad or as cooked greens (Silverman 1977).

The beet was original domesticated as a plant grown for its edible leaves. Farmers likely plucked the leaves off of the young rosette as needed. Some varieties of *Beta vulgaris* ssp. *vulgaris* continue to be used as cooked greens in its various forms, including leaf beet, Swiss chard, and spinach beet. Even though the table beet's leaves (beet greens) are smaller, they can be cooked the same way. The taste is nearly identical (see the chapter 5 recipe). But we use lots of leafy greens. Use as a potherb doesn't do much to justify the unusually high sugar content that Marggraf found in the beet. Let's take a look at the next generation of *Beta vulgaris* ssp. *vulgaris*, the table beet.

Historians of the beet disagree about when beetroot for food evolved from leaf beet. They do agree that roots of the beet were used for medicinal purposes for thousands of years. Some scholars assert confidently that beetroots for human consumption were enjoyed by the classical civilizations of Europe (Biancardi, Panella, and Lewellen 2012). Others are firm in their belief that enlarged sweet roots for the table were established sometime in the medieval era; early Renaissance artwork unquestionably reveals table beets that would be recognized as such today (Francis 2006). The data are thin and hard to interpret. For example, it's not clear whether Romans writing about root vegetables were concerned with the beet or the totally unrelated turnip. Regardless of *when* table beets evolved, here's one plausible, if hypothetical, scenario for *how* it happened:

If you've ever grown lettuce in your garden a little too long, you know how that delicate tasting cluster of leaves suddenly shoots up into a stalk of nasty-tasting ones. The

FIGURE 5.2 Table beet taproots, with cross section and leaves; all parts depicted are edible.

metamorphosis of a rosette plant that bursts forth with a central stalk is called bolting. The stalk eventually starts flowering flowers. For many bolting species, yummy rosette leaves dry up and are replaced by smaller ones on the stalk that are bland, leathery, or downright bad-tasting. For the ancient potherb farmer, it was the end of a useful plant. Therefore, the growers of potherbs tossed the early bolters into their trash piles. Thus, they inadvertently selected for those that delayed flowering and continued to produce rosette leaves. Late-bolting plants ended up mating with one another, their progeny genetically uncontaminated by the long-gone early bolters. Thus, they gave rise to late-bolting descendants, ones that waited until the second year to flower. In short, "the earliest growers selected and reproduced biennial individuals, i.e., those that flowered the year after seeding, allowing for a longer time for leaf production, the only part used for food" (Biancardi et al. 2005).

Delayed flowering gives the plant more time to store more energy (sugar) for the bolting that eventually must take place for sexual reproduction. In the rosette stage, a plant has two substantial parts, leaves and roots. The later a plant flowers, the more time it has for the root to grow. A larger root could store more energy for bolting. We could expect that such plants might sometimes evolve an expanded root, and in the case of *Beta vulgaris*, they did. Substantial taproots are often typical of biennial plants (like carrot, parsnip, salsify) that have rosettes in year one and bolt to flower and fruit in year two. Similar evolutionary events could be suggested for other root crops that spend their first year as rosettes. For example, carrots have significant sucrose content in their roots (Suojala 2000).

With regard to proto-beets, some farmers might have

pulled too hard when harvesting a large handful of leaves from the rosette of their planted potherb for the day's vegetable stew. Finding the expanded root, some curious (or lazy) cook might have thrown it into the stew. Even a little bit of sugar in the root in those sweet-less days could have raised a few eyebrows and motivated a hunt for plants with bigger and more tasty roots. Experimental farmers might have started nibbling on roots and selecting the sweeter-rooted plants. One or more of the beet-potherb lineages began to evolve into biennials. The eventual result was the table beets we know and love today. Beetroots for animal fodder, aka mangolds and mangel-wurzels, might have been selected the same way. The particular mangold variety that is the parent of the sugar beet appears to have been a descendant of a natural mangold × chard cross (Ford-Lloyd 1995). Those beetroots were packed with enough sugar to pre-adapt them for the beet's third evolutionary incarnation as the producer of an industrial chemical.

By the nineteenth century, the study of plant improvement had improved itself into a real science, explaining Napoleon's decree to study how to grow beets for sugar and how to make them into better sugar producers. A half century before Mendel's experiments and a full century removed from the birth of modern genetics, practical agricultural scientists recognized that inheritance worked and could be exploited for plant improvement. The history of the science of plant improvement involves methods with increasingly sophisticated matchmaking. Early on, breeders often practiced (and sometimes still do) a method known as mass selection on short-lived crops that reproduce sexually. The primary difference between mass selection and unintentional selection under domestication is that mass selection is intentional and somewhat organized. Mass selection, domestication, and Darwinian natural selection are fundamentally the same process: evolutionary change occurs because certain genetically based traits increase in frequency over time due to the fact that those traits are associated with a disproportionately higher number of offspring in each generation. Mass selection is a cycle of the creation of diversity by allowing plants to naturally mate followed by subsequent selection from that diversity that continues generation after generation. You just plant them in a field and let nature—that is, open pollination—take its course. Unless plants are obligate outcrossers, open pollination involves some mix of selfing and outcrossing. The more outcrossing the better.

The seedlings are grown in the same field or fields and measured for the trait or traits of interest. Plants representing the best performing individuals are identified. Their seeds are harvested and bulked together. Those seeds are planted. The resulting plants are grown to be characterized and allowed to inter-mate; the next set of progeny are evaluated and selected, and so on. Because the cycle involves a gradual depletion of the genetic diversity originally present in the founding population, the subsequent selected populations may occasionally receive an intentional infusion of more variation by the breeder in the form of planting crop varieties or even wild relatives in the breeding field. Adding variation from novel source populations may be especially helpful if the original base population carried little genetic variation for the trait of interest (Simmonds 1979).

The first steps on the road to building a better sugar beet were those mass-selection cycles. Setting up the mating phase of mass selection is easy for beets. *Beta vulgaris* is a super-outcrosser. Beets are mostly wind-pollinated but are also insect-pollinated. Also, cultivated and wild *Beta vulgaris* are self-incompatible, which, you will remember, means that a plant cannot mate with itself (Biancardi, Panella, and Lewellen 2012; Larsen 1977). Thus, under a mass-selection planting scheme, each flowering beet plant will be surrounded by lots of others of its own kind, getting the opportunity to act as a fertile male for many mates and to bear seed sired by lots of different fathers (Biancardi et al. 2005).

Early sugar beet breeders simply grew their plants in a field plot and let the breeze and the odd insect play Cupid. No artists' brushes transferring pollen; no forced commingling of flowers bound together in a pollination bag. The result was lots of diverse, outcrossed seed, with the genetic contributions of many different parents. This simple way of arranging marriages created bursts of variation that served as a substrate for repeated selection. Mass selection has even been successful for improving sugar beets for certain traits in contemporary times (Biancardi et al. 2005). (Note: Mass selection is *not* the same as the clonal selection that is practiced on long-living plants that are able to be cloned such as banana. Clonal selection is still a kind of human-mediated natural selection, but variation comes from mutation, not sex. Nonetheless, if a breeder can encourage certain banana clones to have sex, then she can practice mass selection on the resulting progeny [Simmonds 1979].)

One shortcoming of mass selection is that the bulking of seed from the best performing individuals may slow the selection process if the excellence of certain maternal plants has nothing to do with inheritance. What if a superior individual simply outperforms others because of environmental variation in the field? For example, what if that individual grew where a dog died (or pooped)? And what if that individual contributes disproportionately more seed to the bulk collection than others whose superior traits are genetically based?

As the 1800s progressed, a refined selection method was introduced for sugar beet improvement: family selection (also known as progeny selection) (Biancardi et al. 2005). A breeder practicing family selection still focuses first on superior mothers that pass on desirable traits. But instead of bulking seed, the breeder keeps the seed produced by each maternal plant separate so that each family of siblings is measured as an individual group and compared with the other families (e.g., each family of seedlings might be planted in its own individual row in the field so that the breeder can examine each family's performance easily). Obviously, the seeds harvested from a plant all share the same mother, even if they probably often have many different fathers. If those progeny do not show the superior performance exhibited by the mother, the entire family of offspring is discarded from the future gene pool. In the case of the biennial sugar beet, they can be removed by the end of the first season so that their pollen does not contribute to the next generation of selected seeds. The results are less messy and more efficient.

Family selection is a scalpel compared to mass selection's jackknife. The breeder still takes advantage of sexual shuffling of the genes, but the method gives the breeder more control over parentage so that gains are not eroded by mating with unintentionally selected ne'er-do-wells. By 1880, less than seventy years after Napoleon's call to action, propelled by cycles of this refined selection method, sugar yields reached 18–20% sucrose per beet fresh weight, more than quadrupling the original value (Francis 2006). Today's percentage fresh-weight yields are about the same (Draycott 2006).

Percentage of sucrose is only one type of yield. Crop breeders are interested in many traits, from disease resistance to the shape and size of the produce to flavor and shelf life. But the most important trait is usually the maximum product per acre. Jonathan Swift ([1726] 1999) said it best in *Gulliver's Travels*: ". . . whoever could make two ears of corn, or two blades of grass, to grow upon a spot of ground where only one grew before, would deserve better of mankind, and do more essential service to his country, than the whole race of politicians put together." About the same time that sugar beets celebrated the anniversary of their first century of genetic improvement, a new crop improvement method—again, based on managed romance—became available for squeezing more sugar out of each acre. To provide some context for the third wave of sugar beet improvement, let's continue the story of the geopolitical adventures of the homely taproot.

The production of sugar from beets has always had its ups and downs. The fall of Napoleon's empire was the first of many setbacks. The blockades evaporated. Cheap tropical sugar from cane returned to Europe. While beet sugar factories were shuttered elsewhere in Europe, the French dug in their heels. Slowly, sugar beet rebounded. Scientific selection paid off. Yields increased. Europe had a new crop. Trade barriers evolved to protect the local industry. The combination of improving the beet and the methodologies for industrial processing, along with economic protectionism, proved so successful that beet sugar began to overshadow cane sugar.

In some cases the success of beet farmers at home jeopardized the profits of the cane plantations in the colonies. A 1901 international agreement to withdraw national subsidies for beet sugar production as well as to remove tariffs on imported cane sugar pushed the pendulum hard in the other direction. Cane sugar became competitive and rebounded to the point that half of the world's sugar production came from cane and the other half from beet. Then World War I and World War II disrupted sea trade; beet sugar rebounded as a domestic alternative, buffering against the extinction of the sugar beet as a crop.

By the end of World War II, early selection methods for short-lived outcrossing crops were overshadowed by a new approach: the creation of hybrid varieties. Remember, commercial hybrid varieties don't have anything to do with hybrids between species or even subspecies. (But we will be getting back to that kind of hybrid by chapter's end.) The first successful commercial hybrid varieties were created for corn. Breeders had long recognized that when they made variety-to-variety crosses by hand, the progeny often had hybrid vigor. Even Charles Darwin ([1859] 1902) commented on hybrid vigor in chapter 4 of his *On the Origin of Species by Means of Natural Selection*: "I have collected so large a body of facts, showing, in accordance with the almost universal belief of breeders, that with animals and plants a cross between different varieties, or between individuals of the same variety but of another strain, gives vigour and fertility to the offspring. . . ." Corn breeders were eager to create intervarietal hybrids. First they needed uniform parental lines that could be crossed to make great hybrids. So, breeders initially concentrated on creating relatively inbred, and thereby relatively uniform, lines of corn. Once these were created, they hand-crossed the inbred lines to see which combination of pairs produced the best hybrids. They found that hybrid progeny of any pair of uniform inbred lines are also highly uniform and, for certain pairs of parental lines, very robust and high yielding. Making these hand-crosses is easier for self-compatible corn than for most crops because corn is monoecious. The male flowers are at the top of the plant in the tassel, and the female flowers are below in the ears, their stigmas exposed as corn silk. That spatial separation of sexes (remember herkogamy from chapter 2?) can be exploited to assure that the only seed produced is the result of intervariety outcrossing.

Once the breeders identified and ramped up their inbred parental lines, they had a new dilemma. How can intervarietal seed be produced in abundance for commercial purposes? Given that corn is wind-pollinated and self-compatible, left to its own devices, it creates about 10% selfed seed and 90% seed from crosses between individuals at varying distances. Even if two varieties were planted as alternating individuals, without human intervention some considerable fraction of natural inter-individual crosses would be the result of mating *within* a variety, and only a fraction would be from the desired *between*-variety crosses. Was there a way to create 100% pure and uniform seed from intervarietal crossing, short of doing it all by hand? In the case of experimental seed, the breeder plays matchmaker. For corn growing either in the greenhouse or the field, the male and female inflorescences are first bagged; then, by carefully timing of unbagging and rebagging, pollen can be strategically transferred by hand from the male flowers of one variety to the female flowers of another. The breeder precisely controls who mates with whom and gets hundreds of hybrid seeds with desired parentage. It is not enough for commercial purposes, but it works well for experiments to find pairs of lines with good "combining ability."

But it's another matter to create millions of seeds to be sold to farmers. Here's how the corn companies first did it: The breeder planted his carefully chosen pair of inbred varieties in a series of alternating rows. For many years, armies of high school and middle school students were hired during their summer vacation to remove ("emasculate") the unopened tassels from one variety. The second variety in the alternate row was left untouched to serve as the only possible local pollen source for the emasculated variety. At the end of the season, the breeder's crew harvested ears from the emasculated variety, each ear bearing only hybrid seed. The hybrid-seed production field is designed with an eye to "arranged marriages." It is a gynodioecious (see table 2.3) population of emasculated male-sterile, but female-fertile plants of variety 1 plus bisexual monoecious plants of variety 2. When the tassels on variety 2 release their pollen, they are the only local plants capable of siring seed on variety 1. Thus, the seed sired on variety 1 by variety 2 is deemed "hybrid" seed because the seed is a result of intervariety hybridization. The non-emasculated variety 2 plants could have mated with anyone; they and their ears were discarded.

Cornfields that grow near these seed-multiplication fields may play havoc with the breeder's matchmaking by cuckolding the chosen father. But years of experience with unintended long-distance pollination eventually taught breeders that 660 feet to the next cornfield provides sufficient sexual isolation, resulting in more than 99% local cross-pollination (Kelly and George 1998). Both those creating the seed and those buying it were comfortable with that tiny level of genetic impurity.

Hybrid varieties were a boon for farmers because they are more uniform and with considerably higher yield than the previous open-pollinated varieties. You already know about the benefits of genetic uniformity to a farmer from prior chapters. Hybrid varieties were also a boon for seed companies because they created a captive audience of customers. Corn farmers who adopt hybrid varieties must buy seed every year. With open-pollinated varieties, farmers saved enough seed from each year's crop to plant the next year's crop. Prior to hybrid seeds, if the farmer wanted to improve her crop, she might buy some seed from one of the seed companies or exchange some of her seed with another farmer down the road. But if farmers saved and planted the seed created by their hybrid crop, the hybrids' genes scrambled in the progeny gave forth a burst of stunted and highly variable plants, as mentioned in chapter 3. Corn farmers didn't mind the shotgun marriage to the seed companies because they were smitten. Hybrid varieties first swept the United States and then the developed world. Once hybrid corn proved a success for the seed industry, breeders rushed to create hybrid varieties for other crops.

For monoecious crops with physically separated male and female flowers like corn, manual emasculation is straightforward. I asked Professor Detlef Bartsch of RWTH Aachen University (who knows as much about the beet and its wild relatives as just about anyone) whether hand emasculation of the tiny hermaphroditic flowers works for sugar beet breeding. He told me that among technicians (who are the heart and hands of almost any scientific project), "good female technicians can emasculate 40–60 flowers per hour. . . . Male technicians—according to my experience—are only half as successful, maybe 20–30 flowers per hour." That's a good number for experimental work by scientists. But emasculation by hand is, to understate the obvious, impractical for large-scale seed production for a plant that produces a single seed per flower! If only beets were gynodioecious species or monoecious species that could be altered to be gynodioecious through simple physical emasculation. Fortunately, sugar beet breeders are clever folks. They came up with an elegant alternate solution: genetic emasculation. They already knew that with the careful examination of countless plants, other plant scientists had found that hundreds of otherwise hermaphroditic or monoecious species produce a few (usually *very* few) male-sterile individuals (Kaul 2012); these variants are often genetically based. With the availability of genetically based male sterility, plant breeders don't need to physically emasculate. They can build a gynodioecious population. With the success of hybrid corn, the hunt for genetically based male sterility was on—not only in beets, but in all kinds of food crops.

American breeder F. V. Owen (1942) discovered and interpreted genetically based male sterility in beets in the 1940s. Specifically, he found a type of non-Mendelian male sterility called cytoplasmic male sterility. In both plants and animals, the vast majority of genes are inherited as pairs on a double set of chromosomes residing in the cell's nucleus. Thus, Mendelian genes are often known as nuclear genes. The chromosomes in the nucleus segregate into single sets during gamete formation. At conception, two single sets join together to create a new double set in the one-celled fertilized egg, the zygote. But a tiny fraction of the genes of plants and animals reside elsewhere, on a single chromosome in another subcellular body or bodies. In the case of animals, the subcellular body is the mitochondrion. For plants, there are two types of bodies that house a single tiny chromosome: the mitochondrion and the chloroplast. Because mitochondria and chloroplasts reside in the cell but outside of the nucleus—that is, in the cytoplasm—their genes are called cytoplasmic. The cytoplasmic chromosomes do not segregate and recombine. Instead, they are inherited asexually; in humans and most plant species, they are passed directly from female parent egg cell to the resulting zygote that develops into the child. In humans, only 37 of their 20,000 genes reside in the maternally inherited mitochondrial chromosome.

In sugar beets, cytoplasmic male sterility (CMS) is inherited as a variant in the mitochondrial chromosome of certain plants. Beets rarely express male sterility. Whether male sterility is expressed is not determined by the CMS variant alone, but by the interaction of the CMS gene and the genotypes of two nuclear Mendelian-segregating genes. Once the inheritance and expression of CMS were determined in the beet, it was clear that the expression of male sterility could be manipulated to create plants that did or did not express male sterility by breeding the appropriate genotypes of the nuclear genes into lines that had the male-sterile cytoplasmic genotype. Simply put, make the right cross, and you can turn on or turn off male sterility in the next generation.

Just like for corn, two sets of preselected inbred lines of sugar beets—one line expressing male sterility and one that was both male and female fertile—could be planted in alternating rows so that the breeze could act to introduce the waiting parties. And also like corn, the resulting hybrid seed grew to be highly uniform and extraordinarily vigorous.

But unlike the original hybrid corn seed, CMS did away with the need for human labor for emasculation. In fact, hybrid corn seed production now works the same way. Hybrid sugar beet seed based on CMS was first introduced for commercial seed production in 1969. The breeder now had even more control in terms of diversity and selection. Pairs of inbred lines were carefully created and chosen to create the optimal sugar beet. With the breeder as matchmaker, good sex had triumphed—or had it?

The Mediterranean Basin may be the most romantic place in the world. Provence, Catalonia, Tunisia, the Greek isles, and Tuscany are the stuff of sultry novels, healthful cookbooks, and cycle touring catalogs. The subtropical climate that attracts tourists is perfect for the herbs and produce that make the various cuisines among the tastiest in the world. The wild ancestors of the woody aromatic flavorings—thyme, rosemary, oregano, bay laurel, and lavender—still grow in the Mediterranean's winter-wet, summer-dry shrublands. The warm golden afternoons create the right conditions for natural air-conditioning. The air lazily rises in the heat of the day. Sometime in the morning, a cool, steady breeze from the sea moves inland, filling the void left by the heated air. The breeze also carries dust of the fine beach sand, tiny crystals that are but memories of salt spray, and pollen from coastal plants (Ellstrand 2003).

For much of that pollen, the journey is a lonely one. Most coastal species hug the sea strand. Pollen blown inland from those species is doomed to death just as sure as being blown out to sea. But pollen of a few species may find a mate far from home. One example is the pollen of sea beets, the wild ancestor of domesticated beets. Although they are tightly restricted to the coasts of northern Europe, sea beets have a more relaxed distribution along the Mediterranean, where they grow both on the coast and sometimes miles away in the human-disturbed habitats in the adjacent inland valleys (Biancardi, Panella, and Lewellen 2012). We already know that beets, wild or not, are predominantly wind-pollinated. For many species in which pollen is transported by the breeze, successful cross-pollination with another plant of the same species can occur at surprising distances, sometimes one or more miles. The beet is no exception. Itinerant sea beet pollen may find a mate among its cousins growing on the roadsides and ditches of Mediterranean inland valleys. Alternatively, the wind may carry both coastal and inland sea beet pollen to fertilize a slightly more distantly related mate: the sugar beet (Biancardi, Panella, and Lewellen 2012).

Seed of the sugar beet is an important agricultural product in Mediterranean regions of southwestern France and northeastern Italy.[1] Much of Europe's multibillion-dollar/euro sugar beet industry depends on that seed.[2] The sea breeze aids in the cross-pollination between the inbred sugar beet lines planted as alternating rows of male-sterile/female-fertile and fully fertile plants arranged to create hybrid seed. As already explained, when physically isolated from other beet pollen by a substantial distance, seeds set by male-sterile plants are productive intervarietal hybrids, designed to combine the best characters of its parents with the bonus of hybrid vigor. The harvested seed is sold to farmers in cooler parts of Europe, such as France, Germany,

1 ESCAA, "Seed Production in the EU," www.escaa.org/index/action/page/id/7/title/seed-production-in-eu.

2 European Commission, "Agriculture and Rural Development," ec.europa.eu/agriculture/sugar/index_en.htm.

Belgium, the UK, and Poland. Isolation is important. European sugar beet breeders recommend that sugar beet seed production fields be isolated by one kilometer (a bit more than a half mile) from the nearest flowering beet to obtain sufficiently high levels of genetic purity (Kelly and George 1998). Imagine a field of hundreds of unharvested table beets abandoned by a bankrupt truck farmer. They would eventually flower, producing enough pollen to contaminate a nearby sugar beet seed production field. A good fraction of the seed produced by the male-sterile seed-producing plants would grow up to be low-sugar, commercially worthless inter-subspecies (sugar beet × table beet) hybrids.

It's important to keep the females from rogues. Consequently, the parts of southern France and Adriatic Italy that were specifically chosen for sugar beet seed production areas are hundreds of kilometers from the major leaf beet, table beet, and sugar beet production areas of Europe. The logic is not unlike that of parents who sometimes send their daughters to isolated girls-only schools that are paired with the appropriate boys-only schools. Indeed, those seed production fields are at least a kilometer from the sea beets that grow in the adjacent region. It turns out, however, that those fields are not so far from the sea beets to fully frustrate long-distance romance. The stray wind-borne pollen grains of sea beets have turned out to have considerably more significance than a stolen kiss. By the time those illicit romances were discovered, they had substantial economic consequences (Ellstrand 2003).

Rosettes of dark green two-foot-long leaves, arranged in neat rows, stretched for as far as the eye could see under the bright blue summer skies of England, the north of France, Belgium, Germany (or the bright blue winter skies of California's Imperial Valley)—sugar beet fields at their best are breathtakingly beautiful. But by the mid-1970s, northwest Europe's sugar beet fields were pocked with tall and spindly plants, spangled first with flowers, and then with seeds. These skeletal weeds were a metaphor for lost profits and economic damage to come. European sugar beet growers received a rude shock. Too many of their beets were bolting prematurely (Longden 1993). By the mid-twentieth century, breeders had worked hard to select for varieties of table beet, fodder beet, and, in particular, sugar beet that bolted extremely rarely, making sure they behaved as biennials. They created well-behaved beets that put up a rosette of leaves and put down a substantial succulent root for harvest. Harvesting follows a straightforward protocol. First, sugar beet leaves are cut off and used for animal food. The root goes off to the processing factory. If left to grow for another year, the cold of winter triggers a lifestyle change. Once spring of the second season arrives, they bolt; that is, these "vernalized" plants send up a flowering stalk, depleting both their leaves, which wither, and their swollen root, which becomes woody and useless. Spent, they set seed and die (Ford-Lloyd 1995). Because sugar beets are harvested during their first season of growth as rosettes, they should never get a chance to bolt.

Prior to the 1970s, the typical sugar beet field had no more than a few individuals that were off on their timing and accidentally bolted. On average, no more than one in a million sugar beets, or even fewer, exhibited such deviant behavior. However, suddenly the fraction of bolters increased. The new renegade beets were living fast and dying young. That is, they behaved as annuals. After a few months as rosettes, they bolted, drawing all of their resources from leaves and root to nourish their flowering stalk and seeds. Imagine the surprise of sugar beet growers from France to Germany whose fields were suddenly infested with hundreds or thousands of bolters. Obviously, with its withered rosette leaves and woody root, a bolter deprives a farmer of any useful product. But a bolting beet is more obnoxious than you might think. Bolters rise above adjacent crop plants, grabbing sunshine, and depressing the yield of their well-behaved neighbors (Longden 1989). Also, their hard woody roots take a toll on farm machinery and the machinery of processing plants. Everyone was willing to put up with the odd bolter, but when they become too common, a field isn't even worth harvesting. And the frequency of these bolters increased in the field year by year (Longden 1993).

Generally, sugar beet bolters do not increase in frequency for three reasons. First, your standard bolter might have some physiological quirk, not a genetic one. Most frequently, an unexpectedly late cold spell stimulates bolting in beets that were planted a bit too early. Thus, if seed from these vernalized bolters drop into the field and germinate the next time a farmer plants a sugar beet crop, the children of vernalized plants should not bolt prematurely. Therefore, the farmer who plants sugar beets every year doesn't have much of a problem if vernalized bolters were present in recent crops because if the seed from those plants germinate at the same time as the crop, they usually grow up to be non-bolters. Second, bolters are typically so infrequent in a field that they have trouble finding a mate. The one-in-a-million bolting plant would occur sporadically in a field, here and there, surrounded by hundreds of thousands of chaste rosettes. They wouldn't get much pollen. Because of their self-incompatibility, they wouldn't set much, if any, seed. Third, the prudent sugar beet farmer knows that bolters should be eradicated and sends out workers to scour the fields and remove the odd bolter by hand. Get that bolter out fast enough, and it will not set seed (Ellstrand 2003).

But Europe's new bolters were different on two accounts. First, they appeared at a relatively high frequency, about 8 of out 100,000 plants, one hundred times more frequent than expected. Mate-finding opportunities increased with the number of bolters. Therefore, they set plenty of seed, as many as 20,000 seeds per plant. Second, their bolting wasn't a physiological quirk; it was genetically determined. Most seeds set by the newly deemed "weed beets" germinated to become bolters. European farmers soon realized that bolters were increasing with each year of sugar beet production as the seed that dropped from the bolters created a seed bank in their fields (Longden 1993). The plague of weed beets established rapidly. By 1981, they infested 3 million acres of sugar beet fields, an area roughly twice the size of Delaware. By the early 1990s, they had spread to become a substantial problem in eastern Europe (Soukup and Holec 2004). Cumulatively, weed beets have caused billions of dollars of problems for Europe's sugar industry (Ellstrand 2003).

Crystalline sugar from both beets and cane is a product whose annual global value exceeds $80 billion (USDA-FAS 2016). Sugar is no longer simply a sweetener for desserts and beverages. Sugar is an ingredient in non-edible products like transparent soaps; furthermore, it is a chemical precursor for other biochemicals, such as resins. Brazilian automobiles run on biofuel mixes derived from cane sugar. The sugar beet remains Europe's primary source of sugar. But it is declining in importance as one of the world's major crops. In 1970, sugar beet was the world's twenty-second most important crop in area planted; by 2014, that ranking had dropped to twenty-ninth. The same time period saw a dramatic decrease in area planted, while most other important crops enjoyed an expansion. Sugarcane has steadily increased in area planted (data and analysis courtesy of FAOSTAT; see chapter 3). Even though part of sugarcane's success versus the sugar beet is due to its contemporary use as an ethanol crop, weed beets have also been a component of that change. As demand for sugar increased worldwide, and sugar beets became increasing costly to grow and process because of intermixed weed beets, sugarcane obtained an advantage. Now, nearly a half century later, the weed beet problem isn't going away. More weed beets than sugar beets—more than 20,000 bolters per acre—are growing on some European farms. For perspective, one study estimated that 2,500 weed beets in an acre of sugar beets represent a yield loss of well over a ton.

Controlling weed beets is a headache, costly and difficult. Weed management must include strategies that both remove weed beets in the field and reduce the number of weed seeds in the seed bank. Because weed beets and sugar beets are the same species, their seedlings and rosettes look the same, until they bolt. Any herbicide (weed-killer) that is harmless to sugar beets has no effect on their weedy cousins. Skillful herbicide application by hand can control weed beets that sprout between the planted rows. But only a few practices work to control weed beets that sprout in the crop rows. If broad-spectrum herbicides are applied prior to planting sugar beet seedlings, then most germinating weeds are killed. For those that germinate and bolt later, a tractor-drawn wick, soaked in herbicide and held high above the crop rosettes, is used to selectively kill the taller bolters that it contacts. Another effective method is to wait for the weeds to bolt and then kill them one at a time through hand- or hoe-weeding, long on human labor and time-consuming. These complex management programs may work to reduce weed beets to acceptable levels, but they aren't cheap or easy (Ellstrand 2003).

At least three years of crop rotation—that is, growing crops other than sugar beets until the weed's seed bank is depleted—is effective because most of Europe's crops can easily outcompete the wimpy weed beet. That program can purge over 98% of the weed beet seed bank. A five-year rotation just about drives them extinct. Shorter periods of rotation do little to deplete the seed bank because only a small fraction of seeds germinate or die in the first couple of years. On the other hand, if growing sugar beets is the primary reason for farming, why grow alternate crops for years?

As weed beets became more common, broader and more important questions emerged: Why were the bolters increasing so rapidly? Where did these plants come from? It soon became apparent that at least some—if not most—weed beets had germinated from seeds sown right out of the bag. Clearly, weed control wasn't going to do the job if bags of sugar beet seed were contaminated with bolter seed. Eradicating weed beets meant identifying how they suddenly started infesting bags of commercial seed. Initially, only two explanations were proposed. One theory was that they were sugar beets gone bad, atavistic mutants that reverted to the annual form. Another theory was that these plants were annual wild sea beets, *B. vulgaris* ssp. *maritima*, the subspecies native to the coasts of Europe and the evolutionary ancestor of all cultivated beets. That is, sea beet seed somehow ended up co-mixed with sugar beet seed. Eventually, a third theory was proposed. The bolters from the bag were the children of philandering sugar beets that had engaged in opportunistic long-distance romance with annual sea beets. In this case, a romance with an inappropriate suitor would indeed be a dangerous liaison. Those supporting the third theory argued that such a tryst couldn't have happened with any appreciable frequency until the production of hybrid seed—that is, until thousands of lonely male-sterile beets, *B. vulgaris* ssp. *vulgaris*, sat in seed production fields waiting for pollen from a mate, more or less indifferent to whether that mate was a male-fertile variety of the same subspecies or a male-fertile individual of the wild annual sea beet. This inter-subspecies hybrid theory was supported by the fact that weed beets only became a substantial headache after the hybrid sugar beet seed production emerged during the 1970s. But the theory was at odds with the prevailing view that pollen from plants that were both physically and genetically more distant would have little chance of affecting pollination when competing with nearby suitors chosen by the breeder, which were, after all, individuals of the same subspecies (Ellstrand 2003).

It was time to use some science to figure out how Europe's weed beets had evolved. As we have already seen in the case of the avocado, without genetic tools, establishing a plant's parenthood is nearly impossible. By the early 1990s, the weed beet problem had become so severe, it was clear that identifying the source of weed beets would provide crucial information for the prevention and management of the scourge of Europe's sugar industry. Genetic detectives got to work on the problem. Various research groups sampled the bolting weed beets and the other non-bolting beets in the field. They genetically analyzed them and compared them with non-bolting cultivars as well as the wild, but non-weedy, sea beet. The first question asked about these children of uncertain parentage was "How did they obtain a gene for bolting?" At that time, the genetic basis of bolting in beets was thought to be controlled by a single gene and that the genetic variant, the *B* allele for bolting the first season, was generally dominant over the variant for delayed flowering. Dominance means that bolting is expressed whether the allele is present in one or two copies. (Note: We know now that it can be a bit more complicated, but at that time and for our purposes at hand, the one-gene model suffices). Sugar beets don't bolt prematurely; they don't have the allele. It is always possible that a mutation could give rise to a new bolting allele. If such a mutation occurred in lines grown for seed production or in a breeding program, it would be expressed instantly in the field, and the offending plant or plants would be eliminated under a breeder's watchful eye. It would be impossible for so many mutants to escape, unless perhaps some environmental trigger for expression became newly present in northern Europe's sugar beet production fields that was absent in research fields in southern Europe's seed production areas.

What about the bolting allele's distribution in wild populations? The *B* allele is common in wild sea beets of the Mediterranean but totally absent from populations along the English Channel and North Sea. This geographic pattern is an important clue because wild sea beet populations growing closest to where weed beets first appeared—that is, regions of sugar beet production—do *not* have the bolting allele. Thus, weed beets could not have grown from seeds that blew into sugar beet fields from northern coasts, sometimes as close as an hour's drive away. In contrast, sea beets growing closest to the sugar beet seed multiplication areas not only have the *B* allele, but they have it in abundance. This pattern alone suggested that the origins of weed beets were more likely to be connected to Mediterranean sea beets (either directly by wild beet seed contamination of commercial lots or indirectly by hybridization) than northern wild sea beets. However, this argument was based only on one line of evidence from a single gene. It was still weakly possible that an extremely rare mutant sea beet from the North Sea coast was finding its way into the sugar beet fields. The hunt for supporting evidence intensified (Ellstrand 2003).

The first research report came from France, the world's foremost sugar beet producer at that time and the country hardest hit by the weeds. In 1992 Santoni and Bervillé of Dijon's Station de Génétique et Amélioration des Plantes were the first to genetically compare French bolters with their possible progenitors. Their data led them to the conclusion that "the presence of bolting plants are due to uncontrolled pollination [of sugar beet] by wild annual beets" (Santoni and Bervillé 1992). That bold conclusion was based on thin data. They had surveyed variation in only a single gene of twenty French bolters from three locations. A more thorough analysis was necessary.

French geneticist Pierre Boudry emerged as the forensics hero. He was hardly a seasoned Sherlock Holmes when he accomplished the bulk of his beet research. He was a graduate student, and this important work was his PhD dissertation project. He and his colleagues conducted a series of comprehensive studies. They amassed a greater variety and larger sample of cultivated beets: thirty-five sugar beet varieties and four table beet varieties. Their wild collections included an array of not only forty-three wild sea beet populations from coastal western Europe (from the North Sea to the Mediterranean), but also three wild beet populations from inland areas in southwestern France. In addition, they sampled nine populations of bolters straight out of northeastern France's sugar beet fields. In all, seed was harvested from two hundred plants. The resulting seedlings were analyzed.

They used a molecular genetic marker that was associated with maternally inherited CMS. The presence of that marker confirmed the ancestry of the male-sterile sugar beet (the seed parent for creating commercial hybrid seed). The CMS marker predominated in about 90% of the bolting weed beets of northeastern France. These data supported those of Santoni and Bervillé (1992), confirming that the maternal ancestor of France's weed beets is the sugar beet (Boudry et al. 1993).

What about the weed beets' paternal history? Boudry's group used six nuclear genes to genetically characterize their samples, but they were unable to find alleles that were specific to domesticated beets. They were, however, able to use a mathematical approach to calculate a set of "genetic distances" (estimates of genetic affinity) for their samples. They found that the genetic constitution of France's weed beets were equally distant and intermediate to the cultivated beets and the wild beets of the Mediterranean. In fact, the bolters of sugar beet fields in northeastern France were genetically closer to the cultivated varieties and the Mediterranean wild beets than they were to geographically adjacent populations of wild beets, just kilometers away on France's north coast. The nuclear data specifically pointed to hybridization between cultivated sugar beet and the sea beets of southern France as the most likely source of the weed beets of northern France.

In a subsequent study, the research group concluded, "Weed beet populations are likely to have originated from accidental pollination of cultivated beets by . . . [wild] annual beets in seed production areas. These hybrid seeds—carrying the CMS cytotype and the *B* allele—were transported and sown in northern sugar production areas. They were able to bolt, flower, and set seed after a few months in sugar beet fields due to the dominance of the annual habit (allele *B*). Short crop rotations and use of chemical herbicides instead of mechanical or manual weeding allowed the maintenance and the development of weed beets via a seed bank" (Boudry et al. 1994). They pointed out that "the first commercial variety carrying the CMS cytotype was released in Europe in 1969" (Boudry et al. 1993) concordant with weed beets becoming a major problem on the continent a few years later. As the century's end approached, the same research group conducted an even larger and more thorough study. The patterns were identical (Desplanque et al. 1999). The cutting-edge technique to create the perfect hybrid sugar beet also gave birth to its greatest enemy. And what of that young scientist Pierre Boudry? The success of this real-life C. Auguste Dupin with wild beets foreshadowed his subsequent stellar career. He did not depart from the world of food organisms. He is presently one of the world's foremost geneticists to work on the genus *Crassostrea*, that is, oysters.

Encouraged by the results of the research in France, a German team led by Detlef Bartsch took on the other important source of European sugar beet seed in northeastern Italy's Po Valley, not far from wild sea beet populations along the Adriatic Sea. This production area provides sugar beet seed for mostly Italy and Germany. They compared the genetic characteristics of seedlings grown from weed beet seed collected from sugar beet fields of northeastern Italy and western Germany with seedlings of pure sugar beets and pure sea beets from northeast Italy's Adriatic coast (Mücher et al. 2000). Using over one hundred DNA-based and visually scored genetic markers, they came to the same conclusion as the French team. Long-distance romances between wild sea beets of the Italian coast and nearby sugar beets grown for seed production generated the bolting beets that were wreaking havoc in Italian and German sugar beet fields.

The origin of the weed beets was established, but how to explain the increasing frequency of weed beets in the fields? An increasing fraction of weeds in commercial seeds or the children and grandchildren from those monster plants? Just after the turn of the twenty-first century, another French research team (Viard, Bernard, and Desplanque 2002) used an even greater number of molecular genetic markers to compare bolters germinating precisely in sugar beet rows (presumably planted that year) with those growing out of place between the planted rows (presumably seed dropped by bolters in years gone by). The bolters in the row had genotypes you would expect of first-generation inter-subspecies hybrids that were planted directly from purchased seed. Genotypes of bolters appearing out of place between the planted rows had the scrambled genes that you would expect from the descendants of first-generation hybrids. These weed beet populations were comprised of the descendants of illegitimate liaisons and constantly augmented by hybrids straight out of the bag. Even if seed companies clean up their act, the second-, third-, fourth-generation, and on plants can mate with each other, every time further enhancing the soil with seed from bolter parents. Clearly, once weed beets flower and set seed in a field, without a long rotation to purge the seed bank, they are there to stay.

In conclusion, the sudden appearance of Europe's bolting weed beets is due to sneaky liaisons of wild and cultivated beets in the commercial seed production fields of France and Italy. If sugar beet seed production continues in regions that are not *highly* isolated from wild beets, bolters will continue to contaminate much of Europe's commercial seed. If sound weed management is not practiced, weed beets will continue to multiply and spread through Europe's sugar beet production regions.

In this chapter we followed the twisting road leading to the evolution of the sweet beet. The end of the road was far from apparent when a scruffy wild seaside plant was first gathered for its tasty rosette leaves. It couldn't have been anticipated back when some descendants of the potherb evolved into plants with fleshy swollen roots suitable for human and animal consumption. Centuries later, a global power struggle catalyzed the emergence of a commodity crop from the edible root. Along that road, various evolutionary branches established and remain to this day—the wild sea beet, leaf beet, fodder beet, table beet, and now free-living weed beet.

The sugar beet's evolutionary origin and progress continued via the best plant improvement techniques of the day: first, mass selection, then family selection, and finally creation of hybrid varieties. There's a newer form of plant improvement that is used to build a better sugar beet that we've neglected in this chapter: genetic engineering. Millions of acres of genetically engineered sugar beets are grown in the United States, carrying a bacterial gene that enables tolerance to the broad-spectrum herbicide glyphosate. Given the contemporary controversy and mystique associated with this technology, genetic engineering deserves its own treatment. And because genetic engineering of plants is yet one more kind of sex, it is the perfect topic for our final major chapter.

A Celebration of Beet Evolution

Here is a tip of the hat to the potherb that made it all happen. The table beet *still* can be used as a potherb. The next time you buy beets at the farmer's market, and you are asked whether you want to keep the greens, say that you do. Don't let the farmer remove them. Take them home with the beets, and cook the leaves just like Swiss chard. Better yet, here's a family reunion of the modest table beet and its multibillion-dollar commodity great-grandnephew:

Start with one quart of fresh washed good-looking beet greens (or supplement your beet greens with enough Swiss chard leaves to get to a quart). If they have been in the fridge for a few days and are wilted, as soon as you chop them from the beets, rehydrate them in a large bowl of cold water, immersing the cut "leaf stem" ends for five to ten minutes until the leaves are no longer flaccid. Botanically, leaf stems are not stems—they are petioles. For instance, most of what we commonly eat of celery is the greatly enlarged petiole of the leaf. If you would rather not mess with the remaining beets (I suggest roasting, skinning, and pickling), start with a quart of Swiss chard leaves instead.

If the petioles of either your Swiss chard or beet tops are tender, keep them; if annoyingly fibrous, remove. If in doubt, experiment, if your audience is game. Chop the greens chiffonade-style.

The remaining ingredients are

1 tablespoon bacon drippings reserved from high-quality bacon
1 tablespoon of your favorite cooking olive oil
1 to 4 cloves of garlic, minced
¼ teaspoon of sea salt
1 teaspoon of white (beet) sugar (of course, you can use cane sugar, but that misses the point)
3 to 4 tablespoons unfiltered apple cider vinegar (please avoid white vinegar)

Heat the bacon fat and olive oil in a twelve-inch iron skillet until hot but not smoking. Add and stir-fry the greens until you note the first sign of softening. Stir in the minced garlic. Stir-fry ten seconds. Sprinkle the sugar, salt, and apple cider vinegar over the mixture, and give a single thorough deft stir. Immediately, place a top tightly on the skillet, and turn off the heat. The residual heat from the skillet will boil the cider, nicely steaming the greens. They should be ready in about three to five minutes. Before serving, stir a bit to coat the greens with the oil, drippings, and vinegar.

Serves two. No kidding. A quart of greens cooks down to a few cups, max.

 *Squash and More*

SEX WITHOUT REPRODUCTION

Rose is a rose is a rose is a rose.
Loveliness extreme.
—Gertrude Stein, *Sacred Emily*

Like twined vines, roses and romance partner. The colors of roses received unsolicited have traditional interpretations: Red indicates love. Pink, appreciation. Orange, enthusiasm. Lavender, enchantment. Yellow, warmth, or—depending on whom you ask—infidelity. Certain colors have been a challenge for rose breeders. "Black" roses are deeply dark red. True black has been elusive. Thus, there is no meaning assigned to the black rose. The mauve to blue to violet range has likewise been frustrating. Roses simply do not have the genes to create the compound delphinidin to create a blue color. But never mind that the blue rose, or anything remotely close to blue, could not be bred; a meaning has been assigned to the color: hope for the impossible love—fitting for a flower that could not exist.

Until recently.

A quarter century ago, Australia's Florigene Flowers teamed up with Japan's brewing/distillery giant Suntory to create the first "blue rose." Genetic engineering made it possible after a decade of research and development. The new plant has the biochemical machinery to produce mauve (not true blue, but closer and previously equally impossible) flowers courtesy of a transgene—a human-inserted "cassette" of genes plucked from a bacterium, a virus, and a couple of ornamental plants (pansy and wishbone), which are unrelated to the rose and unrelated to each other. Presently, the Applause variety is grown only in Japan.

Whether you call them GMOs, genetically engineered plants, GEOs, genetically modified plants, transgenic plants, transformed plants, the products of recombinant DNA technology, or the products of biotechnology—you have heard a *lot* about food from genetically engineered plants. From earlier chapters, you know well that genetic modification has been part of plant domestication and subsequent improvement for millennia. Thus, despite the common use of "genetic modification" and "GM" as synonyms for genetic engineering, they aren't really appropriate because genetic modification describes *all* processes involving genetic change in a lineage over time. "Genetic change over time" is, of course, the definition of biological evolution. Genetic change under domestication and plant improvement are examples of intentional human genetic modification. For our purposes, "genetically engineered" and "transgenic" describe plants that evolved via a subset of the newest human-mediated genetic modification technologies that I call, collectively, "agbiotech" (short for agricultural biotechnology).

The technology developed rapidly. Younger than the first personal computer and a bit older than the Internet, we are less than forty years out since the first plant was successfully genetically engineered. As I write these words, it's been only three decades since the first engineered plants were grown in modest field trials in the United States and France, and barely two decades since the first product of a genetically engineered food plant, the delayed ripening Flavr Savr tomato, showed up on supermarket shelves (Charles 2001). Worldwide, hundreds of millions of acres of engineered crops are now planted annually. While it may sound like hyperbole, the claim that agbiotech is the "fastest adopted crop technology in the world" is probably accurate (James 2015).

Yet a lively controversy continues. Are genetically engineered food plants, on the whole, substantially different from other plants? Is there something radically different about the process that creates them? For decades, some champions of genetic engineering have trumpeted the technology—particularly to investors—as a brand-new, thoroughly modern, unprecedented process. Likewise, some opponents have characterized genetic engineering as wholly removed from the natural world. Whether seeing the technology as savior or Satan, both extremes would agree that recombinant DNA technology is otherworldly, novel, and unique.

Both would be wrong.

Genetic engineering is just another kind of sex. Indeed, it is the oldest kind.

In the beginning, life was sex-free.

The first organisms reproduced asexually. These one-celled proto-organisms simply reproduced by doubling their heredity chemicals (maybe DNA, probably not) and dividing into two single-celled individuals. Back in the day, mutations probably occurred much more frequently than now, because the first organisms had yet to evolve the machinery to buffer against the vagaries of the chemical instability and replication mistakes of the hereditary chemicals of the time. Clonal lineages evolved and diverged as different mutations accumulated in different lines. Unsuccessful genotypes fell by the wayside while successful ones proliferated.

Sex evolved some time during the first billion or so years in the history of life. At first, sex had nothing to do with reproduction. As mentioned in an earlier chapter, evolutionists who study the origin and maintenance of sex consider "sex" and "genetic recombination" to be one and the same, that is, the exchange of genetic information resulting in an individual with a new combination of genes. The original sexual one-celled organisms probably evolved recombination via practices akin to those used today by single-celled microbes—from relatively primitive bacteria to the relatively highly advanced and evolutionarily distant protozoa (compared to bacteria, protozoa are almost close cousins to humans).

One way, and perhaps the first type, of recombination is "transformation," which is accomplished when a cell encounters exogenous DNA in its ambient environment and incorporates some of it into its hereditary machinery. Years before DNA was known to be the heredity chemical, British bacteriologist Frederick Griffith (1928) discovered transformation. The typical biology instructor rushes through the transformation story to get to Watson and Crick's elegant discovery of the structure of DNA by way of the follow-up experiments of Avery, MacLeod, and McCarty. Too bad—the tale of Griffith's research is illuminating in its own right.

Bacterial pneumonia is a nasty disease and is often what takes down the elderly who have been hit by the flu. Griffith was motivated to create a vaccine for pneumonia. At that time, researchers were attempting to make vaccines from dead organisms that elicited an immune response to attack living pathogens of the same species. His initial experiment used two different bacteria of the same species: one that caused the disease, and the other that did not. When injected into a mouse, the virulent type killed the mouse; the benign type was innocuous. No surprise. When Griffith injected the dead bacteria of either genotype in a mouse, nothing happened. Still no surprise. But when both living benign bacteria and dead virulent bacteria were injected at the same time, the mouse died. Extracting living bacteria from that dead mouse and injecting them into another mouse yielded another dead mouse. The living bacteria had somehow obtained heredity information from dead ones ("transformed") and had evolved to become virulent. These unexpected results were not going to help Griffith find a vaccine (Fink 2005).

Some scientists might have been frustrated and repeated the experiments over and over again they arrived at the "correct" answer. Others might have given up and started a new line of research to get to a vaccine. But Griffith was the third kind of scientist, the one who follows up with more experiments to confirm or discredit what he found the first time, and, once the surprising data were repeatedly confirmed, to have the faith to accept the grace of scientific reality. Although he started with one research question, he recognized that something else interesting was going on and was willing to let the data lead rather than giving up or butting his head against a brick wall. No chemist, Griffith interpreted his findings as a biochemical "transforming principle" that transferred hereditary information from one organism to another. In this case via a kind of bacterial necrophilia.

That interpretation smelled like alchemy to some scientists, including biochemical bacteriologist Oswald Avery, who initially dismissed Griffith's work as sloppy. But Griffith was soon vindicated by others, including one of Avery's colleagues. By then Avery was fascinated by the transforming principle and led the team that eventually demonstrated that it was the DNA absorbed by the living bacteria that accounted for the observed genetic change (Avery, MacLeod, and McCarty 1944). Combined with bits and pieces of earlier work, those data laid the groundwork for the eventual acceptance of DNA as the chemical basis of inheritance. The fact that Griffith had also discovered an interesting kind of microbial sex was overshadowed by the DNA brouhaha.

Some single-celled organisms have more sophisticated and directed methods of one-way and two-way gene exchange. Under the right conditions, such organisms cozy up to one another, fusing or building a cytoplasmic bridge. The physical connection enables transmission of genetic material. The conjoining of cells and transfer of genetic information is called conjugation. The sex act starts with two independent organisms. The sex act ends with two independent organisms. But one or both have changed genetically with new genes picked up from the other. Again, sex is accomplished without reproduction.

A third example of recombination without reproduction is "transduction," when a virus moves a chunk of genetic material from one organism to another. Also, a virus might leave a piece of viral genome in an organism. This sex act is relatively new compared to transformation or conjugation. Viruses are always parasitic and sometimes pathogenic. They occupy the netherworld between life and non-life. Often, they are not considered organisms. They are highly ancient, because, as a group, they are capable of infecting all other life. But viruses are probably relatively new because they evolved sometime after the initial origin of life; otherwise they wouldn't have had anything to parasitize. Just like transformation and conjugation, virus-mediated sex involves no associated reproduction.

Scientists call such kinds of sex "lateral gene transfer" or, more frequently, "horizontal gene transfer" (HGT) because the genes move laterally from one organism to another. This genetic transmission contrasts with the more familiar "vertical gene transfer" (VGT) that happened when your parents passed their genes to you (necessarily involving reproduction). HGT provides many microbial species with a significant alternative to mutation as a way to accumulate genetic variation. Interestingly, HGT occasionally occurs between microorganisms that humans consider different species. The initial discovery of interspecies HGT was a bit of a surprise for microbiologists. HGT between quite distantly related microorganisms is now well-known (Syvanen and Kado 2012).

While the rate of between-species bacterial HGT is low, it turns out to be tremendously important to humanity. The abundant examples of the rapid evolution of bacterial resistance to multiple types of antibiotics are often due to HGT both within and between species. The newsworthy "poster child" is methicillin-resistant *Staphylococcus aureus*, related to the scary (and occasionally fatal) MRSA (pronounced "mur-sa") disease of health-care facilities. These bacterial lineages evolved and continue to do so by collecting genes from each other as well as other species, including some from beyond the genus *Staphylococcus.*

The evolutionary scientists had a considerably bigger

shock when they found that HGT had occurred between extraordinarily different organisms, between kingdoms of life, even between single-celled organisms and multi-celled organisms. The initial discovery of inter-kingdom sex involved, of course, plants.

Crown gall disease causes tumors in hundreds, if not thousands, of plant species. The diversity of hosts is stunning. Vulnerable food plants run from hops to hazelnuts, from mango to pink peppercorn, from watermelon to quince. The bacterial species that causes the disease is *Agrobacterium tumefaciens*. In the late 1970s, scientists found that the disease commences when the microbe slips a chunk of its DNA into a plant cell. The DNA then integrates itself into one of the plant cell's chromosomes. The bacterial DNA hijacks the cell, directing it to multiply itself into tumor tissue. At the same time, the transformed cells create specific exotic compounds that the bacteria require to survive and multiply. The intercellular choreography to get the DNA from bacterium into plant cell is similar to bacterium-to-bacterium conjugation. While the disease organism proliferates and the tumor grows, the recombined plant cells do not create flowers; this particular inter-kingdom sex act is an evolutionary dead end from the point of view of the transformed plant DNA.

But that's not true for all HGT events. Stably inherited genes received via cross-kingdom HGT have been detected in both plants and animals. One study (Acuña et al. 2012) was published the same day I started writing this chapter. To understand the study's significance, we need to start with chocolate-covered espresso beans and eventually move on to coffee in trouble.

While many humans drink an extract from roasted coffee

seeds, we rarely intentionally eat those seeds. And nobody eats them by the handful. Although tasty (especially when chocolate-covered) and carrying a good dose of vitamins, coffee beans are not particularly digestible for animals. We aren't talking about the undesired effects of too much caffeine; we are talking about the special chemical composition of the seed's storage compound that comprises the bulk of the bean. The seeds of our edible grains and legumes usually store their energy in the form of starches or oils that are easily digestible. The coffee seed stores a lot of its energy in a relatively rare complex carbohydrate called galactomannan. Animals, including humans, don't produce the enzyme mannanase that digests this big molecule into usable chunks. The human intestinal microbiome, the colonic zoo of zillions of individuals of hundreds of species of bacteria, includes a few that produce mannanase (Nakajima and Matsuura 1997) and can ferment the compound into smaller chemicals. Those of us like me who suffer from lactose intolerance—or who regularly use Beano—already see where I am going. Fermentation goes on all of the time in the human gut. But when bacteria ferment a lot of material that can't be digested by human enzymes, the results can be ugly: flatulence, burping, nausea, cramps, and other undesired effects. No chef has yet offered a substantial menu item based on coffee grounds. Save them for the compost.

Between indigestible galactomannan and caffeine that exists at levels that are toxic to very small animals, the coffee tree's seeds should be well protected from predation. With one exception, that's true. But the exception is why coffee is in trouble. The coffee berry borer beetle is the only animal that consumes coffee's seeds. One of only two major animal pests (the other eats coffee's leaves), it causes losses of more than a half billion dollars a year to this global mega-crop. The beetle lays its eggs in the single seed inside the fleshy coffee fruit. The industry calls the fruit a "berry" or a "cherry." But we the botanical cognoscenti know better. It's a drupe. (Botanical side note: The seed is culinarily termed the coffee "bean"; however, the word "bean" is botanically reserved for members of the bean family. Coffee is the sole major human consumable product harvested from three closely related species in the Rubiaceae family. That family's minor consumables include quinine and the medlar fruit. You are more likely to recognize the name of the family's famous ornamental: the gardenia).

The beetle's eggs hatch into tiny larvae that munch through the seed, destroying it. The beetle produces mannanase, an enzyme that digests galactomannan into simple sugars that the insect can use. The enzyme is absolutely unknown in other insects and does not even occur in the most closely related species, the false berry borer. A collaborative team of researchers from Cenicafé (Colombia's national coffee research center) and Cornell University (Acuña et al. 2012) sequenced the DNA of the beetle's mannanase gene and compared it to mannanase gene sequences from dozens of other species, including plants, animals, fungi, and bacteria. The beetle gene sequence most closely matched the sequences of bacteria. This adaptation contrasts with the crown gall story in that the bacterial gene is not an evolutionary dead end. Somewhere in the not-too-distant mists of evolutionary history, the useful bacterial gene horizontally entered and became incorporated into the beetle's hereditary information in a way that could be passed down to future generations. Because it conferred an advantage, it continues from parent to child just like the other genes that came from the borer beetle's vertically inherited insect heritage.

The story of the coffee berry borer beetle's cross-kingdom recombination (sex) is one of many. All multicellular organisms carry some few genes that one of its ancestors picked up by HGT (Keeling and Palmer 2008). One particularly promiscuous gene kleptomaniac is the tiny aquatic animal *Adineta vaga*. This otherwise 100% asexual, clonally reproducing rotifer is known to hoard dozens, maybe hundreds, of genes accumulated by HGT from plants, fungi, and bacteria (Flot et al. 2013).

Plants are well-known for picking up genes by HGT that have ended up being transferred to future generations via vertical sex. Plant-to-plant, virus-to-plant, fungus-to-plant, plant-to-animal, and bacteria-to-plant gene transfers are all well-documented (Keeling and Palmer 2008). As scientists uncover more examples, HGT is taking on significance as a process that can augment genetic diversity in much the same way that hybridization between closely related species can deliver the raw material for natural selection and subsequent adaptation. One difference between the two is that hybridization, a kind of VGT, mixes equal and full contributions of two individuals while HGT most frequently involves an exchange of a small subset of genetic material between two individuals or just from one individual to another.

A second difference is that hybridization cannot occur between cross-incompatible individuals. Interspecies hybridization in plants is not frequent but is not uncommon. Natural inter-generic crossing is rare, occasionally known in a few families such as grasses and orchids (Stace 1975). Spontaneous interfamily (or more distant) hybridization is impossible. Turnips and rutabagas represent two species in the genus *Brassica*. They can cross to a limited extent. The radish, in the genus *Raphanus*, is in same family, Brassicaceae. Attempts to cross a radish with *Brassica* typically fail but very rarely the cross is successful. When it is successful, the plant is sterile. But it is impossible to cross these root vegetables to root vegetables in other plant families like carrots (celery family, Apiaceae) or sweet potato (morning glory family, Convolvulaceae).

In contrast, HGT knows no bounds; it appears to be able to occur within a species or between *any* species. Successfully integrated and heritable genes via inter-kingdom HGT events cannot be common, because if they were, the tree of life could not be perceived as a branching tree, as it is obviously structured by any measure. If HGT were abundantly successful, the tree of life would be a tangled, disorganized web-like Gordian knot of interconnecting branches. But the tree has been repeatedly confirmed by every kind of evolutionary analysis, including the newest molecular genetic analysis based on gene sequences, sorting nicely into obvious branching Linnaean hierarchies. Even a child can see that cats and dogs are more similar to each other than they are to fish. For multicellular organisms, evolutionarily successful inter-kingdom HGT must be extremely rare over a time scale of years and even centuries. Nonetheless, HGT occurs frequently enough to be detectable and to make an occasional profound difference in certain evolutionary pathways.

To sum up, horizontal gene transfer is a natural sexual process that moves genetic material from one organism to another without the accompaniment of reproduction. The transfer can occur within species, between related species, and between the most genetically distant species possible. Genetic engineering (including the newest set of techniques called gene editing) is a human-mediated process that introduces genetic material into an organism without the accompaniment of reproduction. The transfer can occur within species, between related species, and between the most genetically distant species possible. As a process, genetic engineering is the same as horizontal gene transfer. Just as human-mediated selection is a type of natural selection, genetic engineering is human-mediated HGT. Indeed, when plants are genetically engineered, we call them "transformed" by "recombinant" DNA technology. The vast majority of plants currently carrying engineered genes owe them to one of two processes. The scientists who first accomplished artificial HGT/genetic engineering in plants did so using *Agrobacterium* as the agent for transformation (Charles 2001). Plant genetic engineers use a strain of *A. tumefaciens* that is "disarmed," that is, rendered incapable of causing crown gall disease but still capable of transferring genetic information to plant cells. The tumor-producing DNA, a hula-hoop-shaped "plasmid" that the bacterium naturally inserts into a plant cell, is removed by the scientist. But the DNA to accomplish transformation is left intact in the bacterium. The scientist then replaces the disease-creating plasmid with a crafted artificial plasmid that contains both the gene of choice to do the desired work (e.g., make the plant resistant to a disease-causing virus) as well as all the pieces necessary to get the gene to operate in its new home. The artificial plasmid also includes a gene called a selectable marker to help the scientist sort the successfully transformed plant cell from those in which engineering failed. In the case of the "blue" roses, the selectable marker was a gene for resistance to the antibiotic kanamycin.

The scientist does the matchmaking. The engineered bacteria are introduced to plant cells growing in culture, such as a soup of individual cells in a flask or a film of individual cells on gel in a petri dish. Intercellular romance ensues. The scientist selects for plant cells that have been successfully transformed by killing the ones that have not. In the case of the commonly used kanamycin resistance gene, the scientist simply adds that antibiotic to the cultured cells. Plant cells are not naturally resistant to kanamycin. The untransformed plant cells die. The only surviving cells are those that have the plasmid incorporated into their genetic machinery and are expressing resistance (Ronald and Adamchak 2008). Then it is time to coax those survivors to start growing into plants as we know them. Being able to culture plant cells or tissues and regrow plants from them had been going on for years before genetic engineering. The houseplant industry produces millions and millions of plants every year by these protocols. By 1980, the chemical brew to goose plants out of cells in culture had been worked out especially well for the Solanaceae. No surprise that early genetic engineers enlisted plants in that family; our old friend tomato and its cousin petunia were favorites.

The second engineering process is called particle bombardment, biolistics, or the gene gun method. Much simpler to describe (but not necessarily easier to perform) than *Agrobacterium* transformation, biolistics involves first coating small metal (e.g., gold) pellets with a construct containing the transgene and any necessary attachments, and then blasting them into plant cells. Afterward, the selectable marker is used to identify the successfully transformed cells. The original biolistic experiments actually involved using elements of a real gun but some contemporary biolistic "gene guns" bear more similarity to a pressure cooker than a Smith and Wesson. The new set of gene-editing techniques is causing a stir. These technologies have exotic names like CRISPR, ZFN, and TALEN. Despite their increased accuracy for making genetic change, these techniques are fundamentally within the definition of genetic engineering. The hoopla notwithstanding, most of the plant products of genetic engineering will likely come from *Agrobacterium* transformation or biolistic transformation in the near future because so many created by those methods are already in the precommercialization pipeline.

Once individual plants are grown to maturity from *Agrobacterium*- or biolistic-transformed cells, they must undergo further scrutiny. Genetic engineering is just like any other method of plant improvement with a cycle that includes the creation of genetic variation and novelty followed by selection and evaluation. A common misconception in both public discussion and some scientific policy venues is that transformed plants are taken directly from the lab to the field to be multiplied for commercial seed production. But other steps are necessary before engineered seeds can be sold for commercial production. First, initial evaluation takes place with plants grown in growth chambers and greenhouses to assure that no obvious unintended genetic changes have occurred that adversely affect the quality of the final product. Testing genetically engineered food crops typically involves everything from preliminary vigor measures to a full suite of nutritional and other biochemical analysis. Just like in conventional plant improvement, a tremendous number of individuals don't make the cut. Experiments are also conducted to confirm that the transgene is stably integrated into the genome of the selected plants; that is, it shows typical Mendelian inheritance and expression patterns. The creators of genetically engineered plants for market don't want any unpleasant surprises down the line.

Once the most true-to-type lines expressing the new trait to the satisfaction of the engineer are selected, seed are multiplied in the greenhouse in anticipation of growing the plants in the field. Field trials are necessary to verify both that the trait of interest is actually expressed in the array of environments for which the crop is intended to be grown and that the engineered plants have good yield and a quality product. Before a transgenic plant field trial can be conducted, the planned procedures must undergo a regulatory review. While there remain many countries in which field tests are not permitted at all, at present there's no country in which all genetically engineered plants can be freely grown without the permission of government regulators. In the United States, permission to grow transgenic plants outside of the lab falls under the authority of two agencies: the US Department of Agriculture's (USDA) Animal and Plant Health Inspection Service (APHIS) because it already has purview over "plant pests" and the US Environmental Protection Agency (EPA) because it already has authority to regulate products that protect plants (mostly pesticides). The vast majority of field trial applications in the United States can go through a relatively simple "notification" procedure by APHIS in which the transformed organism, the field experiment, and the measures employed to prevent transgenic seed, pollen, or plants from moving out of the research field are described. If the notification is okay, APHIS simply "acknowledges" the notification and the applicant can proceed. A more complicated APHIS application called a permit is necessary in certain cases of field-testing, for example, plants engineered to create a pharmaceutical compound. A single application for an engineered plant may request field-testing at one location or more, perhaps dozens. The subset of plants engineered to protect against pests must undergo additional regulatory consideration by the EPA.

For the United States alone, nearly 20,000 field-test applications have been approved since the first one in 1987. Keep in mind that a field-test application can be for more than one simultaneous field experiment for the same engineered organism, even in different states. The array of organisms and traits tested is staggering. Want to see the list? USDA-APHIS supports a website[1] that keeps track of such information in searchable databases. If we assume that each approved application actually results in a field test, then controlled field tests have been conducted for more than 150 engineered species (plants, microorganisms, and viruses) in the United States. Here's a representative list: apple, barley, coffee, *Dendrobium* orchid, *Eucalyptus grandis*, false flax (a European oil seed crop, "gold of pleasure"), grape, *Heterorhabditis bacteriophora* (a beneficial nematode that is used in gardening as a protectant against insects), iris, Kentucky bluegrass, lime, marigold, *Nicotiana glauca* (tree tobacco), oat, peppermint, radiata pine, soybean, tobacco mosaic virus (yes, even engineered viruses are field-tested),

1 https://www.aphis.usda.gov/aphis/ourfocus/biotechnology/permits-notifications-petitions/sa_permits/status-update/release-permits.

watermelon, and *Xanthomonas campestris* pv. *vesicatoria* (the bacterium that causes a disease of peppers and tomatoes known as black spot).

A representative list of the hundreds of engineered traits that have approved for US field trails is equally illuminating (USDA-APHIS 2017). If you are tempted to skip this A–Z list, think again. It contains some "wowzers": accelerated ripening, branching reduced, branching increased, CBI (confidential business information), delayed flowering, extended flower life, female-sterile, 50% higher nectar sugar, GFP (green fluorescent protein, based on a gene from a jellyfish; very common), heat tolerance, improved bread-making characteristics, juvenile stage reduced, kanamycin resistance (see above), longer stems, male-sterile, not applicable (say what!?), omega-3 fatty acids produced, pharmaceutical protein produced, Quizalofop (a weed-killer) tolerance, reduce nicotine, starch reduced and seed protein content increased, triggers plant defense responses, vaccine production, western corn root worm resistant, *Xanthomonas campestris* ("black spot" disease bacterium) resistance, yield increased, and ZYMV (zucchini yellow mosaic virus) resistant (more later). The diversity of organism-by-trait combinations is mind-boggling. For more detailed information, enjoy more noodling at the USDA-APHIS website.

Not all genetically engineered crops are regulated within the United States. With regards to APHIS, a product is regulated by its Biotechnology Regulatory Services (APHIS-BRS) only if there is some real relationship of the engineered organism to "a plant health risk." For years, the developers of transgenic products sent products through APHIS-BRS review as a courtesy even if they were created by biolistics and contained no genes that came from plant pests (in which case regulation was not a requirement). That changed in 2011 when the Scotts Miracle-Gro Company sent a letter to APHIS-BRS to inquire about the regulatory status of engineered Kentucky bluegrass created by biolistics and without any genes from plant pests. The bluegrass itself is not on the federal noxious weed list. Scotts Miracle-Gro argued that the product did not pull any regulatory trigger. APHIS-BRS agreed and determined that the transgenic grass was not an article regulated under their jurisdiction. The engineered trait, tolerance to a specific weed-killer, is not pesticidal; thus, the EPA cannot regulate it either (Waltz 2012). The US Food and Drug Administration is also part of the Coordinated Framework for Regulation of Biotechnology for plants engineered in the United States, but obviously turf grasses don't fall under their jurisdiction.

Does that mean that it is impossible to obtain a complete list of genetically engineered plants grown in the United States? No, it doesn't. To relieve their liability burden, the genetic engineers want some sort of BRS determination prior to field release. BRS, in the interest of transparency, maintains a website that lists those "Am I regulated?" requests and determinations.[2] As of late 2017, fewer than sixty determinations had been made. Although most of these ended in a determination that a transgenic product was not a regulated article, it is easy to see from the correspondence that BRS did not always offer a free pass to genetic engineers. To the best of my knowledge, not one of the products given the green light is currently commercially available in the United States. Some are still undergoing field-testing. A few are commercially available elsewhere. Some have already

2 https://www.aphis.usda.gov/aphis/ourfocus/biotechnology/am-i-regulated.

apparently gone on to the vast graveyard of failed agbiotech products.

The ability to obtain field trial information for other countries is idiosyncratic. Some countries, such as Australia and Canada, have easy-to-find, user-friendly databases; others, not so easy or no database at all. A conservative guess would be that the number of field trials of engineered organisms outside of the United States would approximately match the magnitude of the United States. If that assumption is true, then worldwide about 40,000 field trial applications have been approved. Given that so many tests have been conducted on so many products of genetic engineering for a quarter of a century, we might expect the markets would be packed with the products of diverse genetically engineered crops with a rainbow of traits. But we would be wrong. Just about everything regarding the characterization of genetically engineered products available to the public cannot be compressed into a soundbite factoid. Those extreme *pro*s and extreme *anti*s like to pick and choose the facts and tilt them for a story embedded within a forest of exclamation points. If you are interested in what "they don't want you to know" (*whoever "they" might be!*), read on. For example, globally, regulators have allowed roughly 150 different crop-trait combinations the opportunity to go to market (that is, they have "deregulated" them) in a period of a quarter century. That's not a lot considering the thousands of combinations field-tested. But in reality the active number of transgenic plants grown for commercial purposes is much smaller. To illustrate, let's examine just the number of species out there.

Contrary to the frequent sloppy verbiage of many of those writing on the topic, "deregulation" and "commercialization" are not synonymous. In fact, some transgenic crops that were commercialized at one time may no longer be available. While there is no website that tells the tale of all of the deregulated crop-trait combinations deregulated, CropLife International provides an online database (www.biotradestatus.com; with an appropriate disclaimer) that attempts to keep track of the status of deregulated transgenic crops associated with some of the larger agbiotech companies. The deregulated list includes about thirty different plant species (edible and not), but only about half of those, thirteen, are actually in the marketplace somewhere in the world. One species, apple, has joined their ranks just in time for me to add this sentence to my final manuscript. The distribution is uneven, like other aspects of agricultural biotechnology, as we shall soon see. Of crops resulting from recombinant technology, only four contribute substantially to global production. Soybean is the gorilla in the room, accounting for almost half of the total global transgenic acreage devoted to all engineered species. Corn is second. Those two make up more than 70% of the engineered fields. Adding cotton (no. 3) and canola (no. 4) accounts for about 99% of the global acreage. (A note on facts relating to the status of biotech crops globally: above and following, I used the International Service for the Acquisition of Agri-Biotech Applications' ISAAA 51-2015 Brief PowerPoint Slides and Tables and the ISAAA's GM Approval Database as my sources.[3] None of the databases that report the global status of agbiotech crops are perfect, including the ISAAA. However, the global facts that the ISAAA accumulates and reports are the best I have found after spending three decades in this field. For the United

3 www.isaaa.org/resources/publications/briefs/51/pptslides/default.asp; www.isaaa.org/gmapprovaldatabase/default.asp.

States only, the USDA-APHIS website is the best first stop of information.)

Just because the primary acreage is dedicated to so few crop species, you might discount their importance. Not so. For example, if you live in the United States—where more than 90% of the planted soybean, corn, cotton, and canola plants are transgenic—a daily encounter with a product of one of the Big Four is almost inevitable. Roughly 80% of American processed food products contain one or more of Big Four ingredients. To illustrate, I just finished an oats and honey granola bar. Its wrapper tells me that it was made, in part, with canola oil, yellow corn flour, soy flour, and soy lecithin. The sugar also listed might be a product of America's largely genetically engineered sugar beet crop—or a product of its non-transgenic sugarcane crop. Furthermore, at the moment, I am wearing a fair amount of cotton clothing (T-shirt, underwear, socks, jeans). Assuming that the cotton was grown in the United States (or China, where my jeans were constructed), then I'd bet you dollars to donuts that the plants that provided the raw material for the threads were transgenic.

Cotton makes its way into some foods as well. Short fibers from the seed are sometimes processed to improve the texture of ice cream and salad dressing. Crude oil extracted from cotton seed is unsuitable for consumption. But refining reduces cottonseed oil odor and radically reduces the concentration of the chemical gossypol. Gossypol was once suggested for an oral male contraceptive, until it was determined that the effective sperm-killing dose was uncomfortably close to the fatal dose (Waites, Wang, and Griffin 1998). Of the world's sources of edible seed oil, refined cottonseed oil ranked number six production-wise in 2016. Its name often appears in a processed food's ingredients list as one of many possible vegetable oils. Also, anchovies and sardines are sometimes packed in cottonseed oil.

Where are the transgenic crops grown? Here's what we know for 2016: Genetically engineered crops for market covered 185 million hectares (457 million acres) in twenty-seven countries on all six habitable continents. That seems like a lot, but here's the kicker. More than two decades after the marketing of the first commercial transgenic crop, the vast majority of acreage remains confined to five countries on three continents. The two leaders, the United States and Brazil (a distant second), comprise much more than half of that production area. If you add only three more (in decreasing order)—Argentina, Canada, and India—the group accounts for more than 90%. A few countries that used to grow transgenic crops (e.g., Iran, Sweden) have stopped, at least for the moment.

Not only are transgenics largely restricted to a few major crops growing primarily in a few countries, but more than 90% of the transgenic seed sold globally is restricted to a small and decreasing number of companies. Seed companies and biotech companies have been merging, splitting, and name changing for a long time, with the mergers dominating. For the ten years leading up to 2015, things had been relatively quiescent. Monsanto was on top, and four other companies had a significant global presence: Bayer Crop Science, Dow AgroSciences, Pioneer (a DuPont business), and Syngenta. The current agbiotech leader and world's largest seed company, Monsanto, was originally a chemical company that moved deeply into the biological side of agriculture by buying almost twenty seed companies as well as absorbing some agbiotech boutique firms such as Calgene, which created the famous Flavr Savr tomato. As I write these words, American-based Monsanto and the German Bayer Group (that includes Bayer Crop Science) are striving to make a merger work with the goal of creating even a bigger giant. Likewise, Dow AgroSciences and Pioneer have joined as part of a greater Dow-DuPont merger. Swiss-based Syngenta, with its own long and highly complex history, has become part of ChemChina. Unless the wave of globalization hits an unexpected breakwater, by the time you read these words, the bulk of the world's agbiotech business will be in the hands of three mega-companies.

What about non-commercial sources of genetically engineered plants? Regulatory hurdles that are relatively easy for multibillion-dollar companies to leap may be daunting for not-for-profit institutions on more modest budgets. By the end of 2017, the United States had deregulated 123 different crop-trait combinations; of those, 119 were created by for-profit entities. Of the remaining four, three came from universities: herbicide-tolerant flax created at the University of Saskatchewan as well as two separate types of papaya engineered for resistance to papaya ringspot virus by Cornell University and the University of Florida. The other crop is a plum engineered for resistance to the plum pox virus by the USDA Agricultural Research Service. Neither the flax nor the plum is commercially available.

The primary transgenic crops come with even fewer primary transgenic traits. The vast majority of the acreage (far exceeding 90%) is devoted to types that are herbicide tolerant, insect resistant, or a combination of those traits. Engineered plants bearing multiple transgenes are often referred to as "stacked" or "pyramided"; that's agbiotech lingo for plants that carry more than one transgene-based trait. Unstacked herbicide tolerance (HT) is the more common of the two significant agbiotech traits, responsible for about half of the world's area devoted to transgenic crops. All the world's genetically engineered alfalfa, sugar beet, and canola fields are comprised of plants protected from one of two broad-spectrum herbicides. Commercial transgenic varieties of corn, soybean, and cotton are either herbicide tolerant, insect resistant, or stacked for both traits. But even for those species, HT predominates.

Herbicides (aka weed killers) have been around for decades, but their application exploded dramatically in the last half of the twentieth century. Originally, broad-spectrum herbicides were rare. Herbicides were generally toxic to one group of plants, but not to others. A farmer growing soybeans might find an herbicide that would kill all of his weeds that were grasses. Using one or more alternate herbicides would be necessary to kill the weeds that were not grasses. Likewise, different crops are naturally resistant to different kinds of herbicides. Genetic engineering offered a tantalizing possibility. What if a plant could be engineered to tolerate an herbicide that killed *all* other plants? That dream was first realized for the herbicide glyphosate (the active ingredient in Monsanto's Roundup). Glyphosate was already known to have the unique property of killing all plants, typically at remarkably low doses, while at the same time having a very low toxicity to humans and other higher vertebrates (it was originally proposed as a flame-retardant for baby pajamas). Another bonus, compared to some earlier popular herbicides, is that glyphosate tends to break down into innocuous compounds relatively rapidly. Given that glyphosate was good at clearing a field of all plants, a farmer growing a crop engineered for glyphosate tolerance could enjoy a tidy, weed-free field (that is, until weeds eventually evolve tolerance for the herbicide). Monsanto scientists discovered the first glyphosate-tolerance gene in a strain of *A. tumefaciens*. Given that Monsanto had already patented glyphosate for use as an herbicide, the rest, as they say, is history. Glyphosate tolerance is the world's number one commercial-engineered crop trait, either stacked or as a stand-alone trait.

Most engineered insect resistance (IR) is based on a pesticide already used by organic farmers. Different strains of the soil bacterium *Bacillus thuringiensis* (Bt) have been found to carry different proteins that are toxic to relatively specific groups of insects. In contrast with broad-spectrum insecticides that have the potential for killing any insect, a specific Bt protein is toxic to only a subset of insects, such as moths and butterflies (Lepidoptera) or beetles (Coleoptera). Each group may contain hundreds of thousands of described and undescribed species, but traditional insecticides often can kill millions of insect and non-insect species. Also, the Bt protein has no known human health effects. Some farmers, both past and present, spray either dead Bt bacteria or bacterial extracts on their plants. A natural insecticide, Bt is appropriate for organic growers.

The bulk of IR-engineered crops (mostly Bt corn and Bt cotton) have a transgene that incorporates a version (frequently tweaked) of a gene that creates Bt's Lepidoptera-killing proteins. The plant carries its own pesticide and expresses it in the tissues most likely to be consumed by the target insect. Bt cotton is one of plant genetic engineering's success stories, particularly in China and India, where expression of pesticide in tissues of the crop has resulted in the dramatic drop of the use of other insecticides that have been externally applied and often pose substantial human health risks (National Academies of Sciences, Engineering, and Medicine 2016). Because cotton has pests that are not moths or butterflies, some pesticides still must be applied. Nonetheless, the annual number of pesticide-associated deaths has dramatically decreased in China and India. Bt crops cannot be used by organic growers in the United States because, as legally defined, "organic" production methods exclude growing transgenic crops.

Compared with the overwhelming use of glyphosate tolerance and lepidopteran resistance, the use of other transgenic traits remains modest. One example is Syngenta's Enogen corn. HT and IR have been designed to help the farmer. But this corn has been engineered with a value-added trait for the biofuel industry. In Enogen corn, the transgene creates a "designer" enzyme that breaks down the starch in corn seed in a way that makes production of the biofuel ethanol simpler and less expensive. With the engineered enzyme expressed in corn seed, an entire step in the industrial production of corn-based ethanol is bypassed. Additionally, the process becomes considerably "greener," saving water and energy. The example of Enogen corn reveals that while the process of genetic engineering may be age-old, the sorts of products can be extraordinarily novel. The primary difference between the process of genetic engineering and prior plant improvement processes is more one of degree than a black-and-white dualism. Genetic engineering can create certain types of products more rapidly than before. But this is not to say that it would have been impossible for more traditional plant breeders to create a similar type of corn for biofuel. It just might have taken them longer.

Or maybe not. Here's a case of a transgenic product that hit the market at the same time as its non-transgenic counterpart. Squashes genetically engineered for resistance to two particularly nasty virus pests were commercialized in 1994; the same year that traditionally improved squashes resistant to the same viruses went on sale. As mentioned earlier, very few deregulated crops have been engineered to be resistant to disease-causing viruses. In addition to papaya and plum, there's one more. Transgenic virus-resistant summer squash occupies more acres than the others put together. As opposed to the Big Four, engineered squash varieties have not eclipsed non-engineered varieties. For years, they have been sufficiently successful to coexist in the marketplace with non-transgenic squashes, even coexisting with its non-transgenic virus-resistant competition.

Summer squash is our chapter's featured crop. Its story will bring us full circle from HGT back to the more conventional (if newer) type of sex. Summer squash is not particularly well-known as an engineered crop. Since it is an underdog, its story is appealing. Approved to be grown only in the United States, I could find no mention of it in any database tracking statistics of engineered crop production and of farmer adoption of transgenic crops in the United States. It barely gets mentioned in the annual ISAAA reports.

Squash is the oldest continuously commercially available engineered crop species. In the early 1990s, China's virus-resistant tobacco was the first commercialized plant transgenic product. It is long gone. Soon after, the Flavr Savr tomato was the first commercialized transgenic food crop but was also pulled from market shelves only a few years after its introduction. Contrary to urban myth, there are no genetically engineered tomatoes or tomato products in our stores. Regarding Flavr Savr's colorful rise and fall, I recommend its biography by Belinda Martineau, *First Fruit* (2001). Those early products aren't the only engineered crops that met an early demise. The museum of biotech dinosaurs includes other crops that were eventually taken off the market (e.g., low-nicotine tobacco)—as well as crops that were deregulated long ago but never put to use (e.g., male-sterile radicchio, herbicide-tolerant rice). Of course, it is always possible that the dinosaurs might be resurrected. Virus-resistant squash has persisted since its deregulation in 1994.

What we call "squash" is a bit complicated. Three different squash species—all in the genus *Cucurbita*—are globally important "squash." *Cucurbita pepo* is the only species that has transgenic varieties for virus resistance. No other transgenic squashes are commercially available, and none are in the pipeline. This annual crop has been genetic modified by humans for millennia. Summer squash might be the first domesticated New World crop, appearing as such in the archaeological record about 10,000 years ago. Almost all summer squash types are *Cucurbita pepo.* Summer squash types are usually cooked when the fruit is still immature and the skin is tender. You know them as the ones whose undeveloped seeds are tiny and soft when you eat them: crookneck, pattypan, and so on. They are also the varieties whose blossoms can be used for cooking—fried, baked, stuffed, in soup, et cetera. But not all *C. pepo* are summer squashes.

Winter squashes are harvested when the seeds are large, hard, and mature, and the skin is tough. Some winter squashes, such as acorn squash and spaghetti squash, are *C. pepo.* Butternut squash is *Cucurbita moschata.* Butternut is my favorite winter squash. Try it slow roasted in a three-way mixture of olive oil, bacon drippings, and butter, drizzled with maple syrup, and dusted with ground dried thyme.

The third species is *C. maxima*. Buttercup squash and hubbard squash are the best known *C. maxima* winter squashes. But don't get too comfy with that simple taxonomy. Each species is incredibly variable, resulting from thousands of years of genetic meddling at the hands of folks as creative as those who gave us Irish wolfhounds, shih tzus, and Labrador retrievers. Consider the pumpkin. Depending on the specific variety, your jack-o'-lantern might be a gigantic type of *C. pepo* or *C. moschata* or *C. maxima*. By the way, the largest fruits of any angiosperm are pumpkins. In 2016 the *Washington Post* (Barron 2016) reported that Belgian farmer Mathias Willemijns set a world record "with a super squash that weighed 2,624.6 pounds" (1,190 kilos). For comparison, that's about a hundred pounds shy of a Toyota Corolla or approximately the weight of *two* male moose. Assuming a 92% yield of pumpkin flesh by weight after removing the peel and the seeds, that big gal could provide the substrate for more than twenty-six hundred pumpkin pies. Grocery store pumpkins are mostly *C. pepo*, and the champion pumpkins are (good guess!) *C. maxima*.

The family that houses pumpkins and squashes is rich with edibles and other useful plants. You probably have guessed that cucumbers and the various gourds are also members of the Cucurbitaceae family. But so, too, are watermelons and other sweet melons, the chayote—a single-seeded fruit that is Mesoamerica's staple "vegetable"—and the luffa (the plant sponge). Bitter melon, until recently virtually unknown to Americans, is increasingly found at farmers' markets and Asian grocery stores. Horned melon looks and tastes like it came from outer space, but it is native to Africa.

Cucurbitaceae is a tight family. Once you get to know one of them, you'll be able to recognize just about all of the

FIGURE 6.1 Male (*left*) and female (*right*) summer squash flowers. Frontal views of both corollas show the five fused petals. In the background are parts of the plant with senescent flowers, and one attached to an already expanding zucchini.

almost one thousand species. They are hairy-leafed, climbing or rambling plants, either true herbaceous vines or woody ones (properly called lianas). Like the Solanaceae, the family is characterized by some interesting chemical compounds. Most notable are the cucurbitacins, which are among the bitterest naturally produced chemicals known. Cucurbitaceous plants have unisexual flowers; thus, the species are dioecious or monoecious. The flowers are typically larger than the ones encountered in previous chapters. Both male and female flowers have five green leafy sepals overshadowed by five substantial, often yellow, petals that are typically fused to one another. Depending on the species, the male flowers have from one to five stamens that are united in some way. For cucurbitaceous female flowers, the part of the gynoecium that becomes the fruit is situated beneath the perianth. Such "inferior ovaries" are rare among flowering plant families (but see banana). If you obtain a young zucchini with the blossom attached, you can easily see how the fruit hangs off the bottom of the flower. The fruits are typically three-carpel berries. Next time you slice a cucumber crosswise, look for the three-ness in the arrangement of the seeds. The fruits often have a hard rind at maturity. Botanists have decided to call berries with a hard rind "pepos." It's okay because *pepo* is the Latin word for a kind of melon.

Summer squash flowers are archetypal for the Cucurbitaceae. Both their male and female flowers bear flashy five-lobed orange-yellow corollas shaped like a floppy baby's hat, set above an inconspicuous green calyx. The fused petals create a corolla that can be four inches long. The stamens prominently displayed in the middle of the male flower are twisted together in a single cone-shaped mass. (Use the males for battered and fried squash blossoms, yum!) The middle of the female flowers features three ear-shaped stigmatic lobes that wait to receive pollen. With flowers so showy and big, you know the plants are animal-pollinated—in this case, insect-pollinated. The original primary pollinators of domesticated *C. pepo* are squash bees, a group of New World pollen-harvesters that specialize on the family's flowers. These bumblebee-sized beauties often still do the job, and they do it well. Domesticated honeybees of the Old World can affect pollination but are inferior. Research has shown that when squash bees and honeybees are both available, the majority of the pepos on a plant are the result of pollination

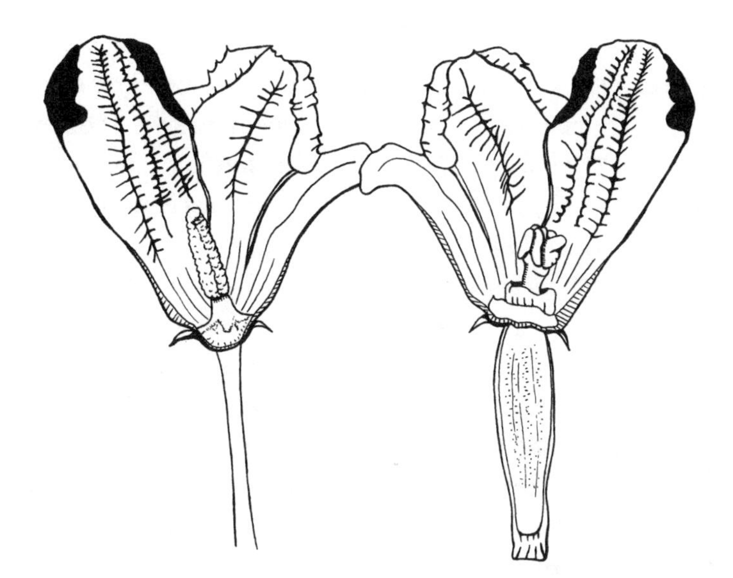

FIGURE 6.2 Cutaway side views of mature male (*left*) and female (*right*) summer squash flowers with exposed stamens and stigmas (respectively) and tiny sepals. The female flower's inferior fruit is expanding into a zucchini.

by the wild native. Other pollinators include bumblebees and cucumber beetles.

As noted above, cultivated *C. pepo* has amazing genetic variation. In addition to acorn, crookneck, and some pumpkins that are hard to tell from those of *C. maxima*, the species includes what is known as zucchini in the United States and Italy and as courgette in England and France. Globally, the latter may be the most familiar form of *C. pepo*. The majority of this species is eaten as fruits harvested long before the seeds mature.

That's an important point. During my grad school days, the steamy summer night potluck dinners featured various forms of summer squash offered up by zoology students lucky enough to have a garden. The zucchini bread was typically edible and often tasty. Other *Cucurbita* donations were not. The problem was that zoology students were not botany students. The potluck zucchinis were mature and mammoth. They approached the size of third-grader baseball bats and were almost as difficult to eat. Steamed, sautéed, baked, or roasted, thick slices made their way into casseroles, lasagnas, mélanges, and even salads. The chunks were so watery that they could not absorb sauce, whether made from backyard tomatoes or American cheese. Fortunately, the blandness was hardly noticeable as we struggled to chew through the balloon-textured skin and the hard dime-sized seeds. After a couple summers, I developed a negative Pavlovian response for anything long and green-skinned with a dusting of light yellow spots. If hunger persisted after a potluck, a stop at the convenience store provided the cure: a Hershey's with almonds, a bag of sour cream and onion Lay's, and a can of 7 Up. Years of tender culinary therapy by my loving wife (a botanist)—featuring slow-cooked, well-sauced, and appropriately immature summer squashes—finally cured my general aversion to *C. pepo*. I have since learned that monster pepos are prized by some in Britain as "vegetable marrows," which are de-seeded, filled with flavorful stuffing, and slow-cooked to infuse flavor back into the flesh. But, for me, a mature baseball bat-sized courgette remains a culinary phobia.

Just as horrifying as an inedible mature summer squash may be to me, equally scary is a virus-infected plant to a farmer of any of the cucurbitaceous crops. An afflicted plant has deformed leaves and flowers. The fruits are stunted and deformed. Yellow spots occur on the green-fruited varieties, while green ones form on the yellow. Consumers won't buy them. Plant yield suffers. The worst possible scenario is grim, complete crop failure. The plagues are multiple: cucumber mosaic virus (CMV), papaya ringspot virus (PRSV), watermelon mosaic virus 2 (WMV2), and zucchini yellow mosaic virus (ZYMV) are the more common culprits.

In the early 1990s, the time was ripe for creating a transgenic virus-resistant squash. Research during the 1980s had uncovered a potential silver bullet for virus resistance. To understand how the bullet works, we need to understand a bit of virus biology. A virus is simple; it is composed of heritable material protected by a coat of protein. The protein coat is largely composed of identical subunits that fit together like LEGO pieces into a chainmail-like tunic. That's all. Building resistance to certain disease-causing plant viruses turned out to be relatively easy. Plant pathologists had discovered that living cells already occupied by viruses sometimes have immunity to further viral invasion. For example, they learned that if plants were treated with an innocuous virus closely related to one that causes disease, those plants sometimes became immune to the nasty one. The scientists reasoned that if plant cells were genetically engineered to express an innocuous virus coat protein at low levels, they should enjoy some resistance. They were right (Beachy, Loesch-Fries, and Tumer 1990). In experiments involving an array of different plants and genes from their disease-causing viruses, scientists transformed plant cells to integrate and express an individual virus coat protein gene. The transformed plants expressed the gene, creating a low level of coat protein within their cells. Such plants were partially or fully immune to infection by the same virus that donated the gene. Note that such a transgene can build reproductively sterile coat protein subunits, but it cannot build the rest of the virus. The logical consequence of the research was to create a disease-resistant plant for farmers. Scientists knew coat proteins don't pose a human health hazard because previous surveys of good-looking produce on supermarket shelves had shown fruits and vegetables often contain detectable amounts of fully operating plant viruses at levels higher than the coat pieces expressed in the transgenics. Furthermore, thousands of generations of hungry humans have consumed less-than-perfect stunted and deformed viral-diseased fruits and vegetables with no ill effects.

In 1992 Upjohn/Asgrow petitioned APHIS-BRS to deregulate ZW-20 a *C. pepo* variety engineered to resist both WMV2 and ZYMV using a transgene expressing coat protein genes from each virus via *Agrobacterium*-meditated transformation. As noted above, *Agrobacterium* transformation is a sufficient trigger for APHIS-BRS to regulate a plant. APHIS-BRS consideration regarding deregulation (making the transgenic plant no longer a regulated article) is essentially a judgment whether engineering has altered the crop in such a way that it is more likely to be "plant pest" than a traditionally created, nonregulated crop. (Nowadays, that includes an environmental impact statement.) If not, then the engineered crop-transgene combination is relieved of BRS regulation. No petition for deregulation has ever been denied by BRS, but dozens have been withdrawn (USDA-APHIS 2017); many of those were probably in response to a request for information from APHIS-BRS (perhaps suggesting—between the lines—that approval would be difficult). But this conclusion is pure speculation on my part. (Note: APHIS-BRS has denied *hundreds* of field-test applications.)

Every APHIS-BRS consideration of deregulation is accompanied by a request for public comment. The Union of Concerned Scientists (UCS), a nonprofit science advocacy organization, brought the squash petition for deregulation to my attention. UCS had previously sent me the Flavr Savr petition for comment regarding whether it might be a plant pest. After reading the petition, I didn't see how it could be. The slowed ripening of Flavr Savr tomatoes was an alternative to picking immature tomatoes that never ripen properly. Its transgene would allow the fruits to be picked mature, packed, transported hundreds of miles, stored, and unpacked onto supermarket shelves just as they were beginning to ripen perfectly. The very best November tomato I have ever tasted was a Flavr Savr (under its market name of MacGregor's). Using genetic engineering to solve a problem for the greater good (or at least for the consumer) seemed like a wise decision for the first transgenic crop to reach the regulators—and the attention of the media. I said as much to UCS.

UCS had chosen me to study the ZW-20 petition because I had already published some scholarly articles about the potential risks of engineered genes entering natural populations unintended (e.g., Ellstrand and Hoffman 1990). My prior research on plant sex and hybridization made me aware that natural cross-pollination between an engineered crop and a wild or weedy relative might introduce a gene into a natural population, possibly resulting in the evolution of a new weed or other problematic plant. Natural hybridization is known to occur between crops and their wild relatives. It varies a lot depending on the crop species involved, but the movement of crop genes into natural populations via standard sex occurs much more frequently than plant-to-plant HGT. Just as in the case of the rise of the weed beet, such hybridization, on occasion, has resulted in the evolution of a new nasty plant. Such problems aren't common, but a few are humdingers. Some examples of unplanned sex between cultivated plants and their wild relations leading to the evolution of problem plants are featured in table 6.1. Not one of those examples features an engineered crop (yet), but they provide a precedent for caution. (Also, a modest number of crops have evolved on their own—without natural hybridization—becoming nasty weeds or invasives, an interesting topic in its own right, but one that we won't follow here as sex doesn't play a key role in those stories.)

The following questions provide a method for judging whether hybridization might create a risk worth examining more closely: Does the engineered crop have a wild or weedy relative growing in the same region as where it will be planted? Are the crop and its wild relative likely to cross-pollinate and create hybrids under natural conditions—that is, are they cross-compatible? Does the hybrid show any significant fertility? If the engineered gene successfully enters the population of the wild/weedy relative, does it have the potential for conferring some advantage to enable that plant to survive better or create more seeds? A "no" answer for any question would be sufficient to conclude that transgene transfer via sex would either be too rare to worry about or would have little consequence in altering the ecology of the unmanaged species. After all, mutations, natural hybridization, and HGT are continuously introducing new genes into natural populations at very low rates, and we don't worry about those natural processes. Even if the answer to all the questions was "yes," it would not necessarily mean that a new weed or invasive would evolve, but it would indicate that more research should be done prior to deregulation (Rissler and Mellon 1996; Hokanson et al. 2016).

TABLE 6.1. *"Bad Seeds": Problematic plants that evolved from unplanned sex between cultivated plants and their wild or weedy relations, some examples**

Sex partners	Location / time of tryst	Problem child / where?
sugar beet × sea beet	southwestern France / repeatedly, late 20th century	weed beet / Europe
sugar beet × sea beet	northeast Italy / repeatedly, late 20th century	weed beet / Europe
cultivated rice × wild rice	Asia / 19th century or earlier	weedy rice / southeastern United States
cultivated sorghum × Johnson grass, intentionally introduced range grass from Old World	southeastern United States / 19th century	Johnson grass with increased weediness / North America
cultivated sorghum × wild *Sorghum propinquum*, accidentally introduced Old World weed	Argentina / unknown	Columbus grass / Australia, North America, South America
cultivated pearl millet × wild pearl millet	Africa / continuing	weedy pearl millet / Africa
cultivated artichoke × wild artichoke thistle, accidentally introduced European weed	California / late 19th century or early 20th century	some populations of invasive wild artichoke thistles / California
cultivated radish × jointed charlock, accidently introduced Old World weed	California / late 19th century	California wild radish (both a weed and an invasive) / Pacific North America

*Adapted from Ellstrand et al. 2010.

With regard to the UCS request to consider environmental downsides of Flavr Savr, the tomato has few wild or weedy cross-compatible wild relatives in the United States. The few present are very distant (hundreds of miles) from commercial tomato-growing regions. Furthermore, I spent days thinking about how delayed ripening could possibly effect the ecology of a weedy tomato so as to transform it into a "Killer Tomato." Other evolutionary and ecological geneticists did so as well. Most of us didn't see any reason to offer a cautionary opinion on the new tomato. The vast majority of the opinions offered against the deregulation of the tomato generally had little to do with the ecology of the organism.

The situation regarding transgenic virus-resistant squash was the opposite of the tomato. Here was a trait that might help a plant become a weed or invasive by fending off some natural enemies. After all, the newborn field of invasion biology had accumulated a lot of examples of species that were benign in their homelands only to turn into nasty invasives when introduced into new lands and freed from their biological enemies—from bunnies in Australia to St. John's wort in the Pacific Northwest. (Remember the Red Queen in chapter 3?)

Did summer squash have any wild relatives around? Were they weeds? If so, could they easily cross with *C. pepo*? If so, were the hybrids fertile? If so, was there evidence for whether the transgene would actually increase weediness? The Upjohn/Asgrow petition addressed these questions. The petition is still available online.[4] The company's conclusions were straightforward: In the United States, *Cucurbita pepo* naturally hybridizes with two wild squashes, *C. texana* of

4 www.aphis.usda.gov/brs/aphisdocs/92_20401p.pdf.

Texas and *C. pepo* variety *ovifera* of the southeastern United States. No mention of weediness is made for the first, and with regard to the second, the petition states: "*Cucurbita pepo* has never been reported as a weed." Crop-wild hybrids are known to be as fertile as the parents. In an experiment comparing hybrids with and without the transgene, no difference in vigor was observed. The petition noted that the virus was not present in the experiment and that more experiments were under way. Furthermore, the petition is sanguine about keeping the gene down on the farm. It notes that seed dispersal is unlikely because squashes are harvested when immature. While unharvested fruits may be plowed into the soil, with mature seeds germinating the following spring, volunteer squash "are eradicated by standard tillage practices." Likewise, the petition discounts opportunities for the crop to mate with wild populations:

> *C. texana*'s geographical distribution is restricted and does not include Georgia or Florida, where the majority of yellow crookneck squash are produced. Therefore, geographical overlap of these two species is limited. In areas of overlap, transmission of genetic material to *C. texana* should be further limited by the fact that squash pollen only travels short distances. For example, a distance of 400 meters serves to isolate genetically engineered squash for seed production....

Finally, with regard to the possibility that adding virus resistance to wild squash might contribute to increased weediness, the petition provided two arguments why a gene for virus resistance should not make an evolutionary difference in a natural population. The petition uses another species, *C. foetidissima*, as an example of a wild species that already has virus resistance and "is not known to be a significant weed problem." Furthermore, in a follow-up study to an earlier petition, Upjohn/Asgrow located fourteen natural populations of *C. pepo* variety *ovifera* and examined an individual plant in each one for visual symptoms of virus infection (nine of these populations were sampled only once). The individuals—and, by extrapolation, the populations—were judged to be free of viral infection. They concluded that virus resistance could not help a plant that suffered no viral disease.

Because the squash was the first transgenic crop that had a trait that might confer an advantage to a wild relative, lots of academic plant scientists—from virus experts to weed scientists to taxonomists to ecologists to evolutionists like myself—around the country examined the petition. They did some fact-checking. Keep in mind that in the early 1990s Internet searches were limited, and that both the Upjohn/Asgrow scientists who wrote the petition and the scientists who critiqued it were able to access only a modest piece of the ever-ballooning elephant that is the scientific literature. For example, some academics might have thought twice if they had known that traditionally improved virus-resistant squashes were making their way to market at the same time as the transgenics. Nonetheless, the academics encountered some pieces of the elephant not discovered by the corporate scientists.

To get a definitive answer, APHIS-BRS commissioned a report on the potential risks posed by gene flow from that crop to its wild and weedy relatives. The world's expert on wild and weedy squashes in North America, Texas A & M's Professor Hugh Wilson, laid out the informed state of the art regarding wild and weedy *C. pepo* in the United States at that time (Wilson 1993). First, Wilson explained that *C. texana* is the same species as any other free-living *C. pepo* (henceforth lumped as "free-living *Cucurbita pepo*," or FLCP). He corrects other mistakes in the petition. The range of FLCP does indeed overlap with summer squash because the vegetable is a favorite of home gardeners. Contrary to the petition's optimism regarding weediness, Wilson notes that FLCP was recognized as a weed problem in Arkansas, Louisiana, Mississippi, Illinois, and Kentucky. In fact, it had been listed as one of the top ten weeds of Arkansas for several years. His report is not entirely pessimistic. Wilson points out that a number of chemical herbicides had been effective at controlling FLCP outbreaks.

More admonitions followed in the report. With regards to crop-to-weed gene flow, he describes an experiment that he had published five years earlier documenting surprising levels of successful natural cross-pollination from domesticated squash to FLCP at a distance of 1,300 meters (0.8 mile!), far beyond the petition's presumed 400-meter isolation distance. Also, he found that crop-weed hybrids displayed hybrid vigor compared to their parents, suggesting that crop genes, transgenic or not, would be boosted into weed populations. Wilson's report concluded with a suggested experiment to test how much of an advantage the transgenes might confer if they ended up in a wild or weedy population.

Professor Wilson wasn't the only one with reservations. APHIS received a lot of feedback suggesting further experimental and descriptive work to address whether FLCP populations were kept at bay by the viruses in question and/or whether the transgene might make a difference in their weediness. I was one of many scientists who called for more experiments with a four-page single-spaced letter (and probably too much attitude) that concluded:

> If (. . . a big unknown) the relevant viruses regulate populations of FLCP, and if (an equally big unknown) the virus resistance transgenes confer a fitness advantage in FLCP, then the transgenes would "tip the balance" and make a marginal weed an economic and environmental problem.
>
> In a sense, determination of nonregulation of ZW-20 squash might be the ideal ecological experiment. If enhanced weediness of North American FLCP occurs by gene flow from transgenic squash, the cost, at most, would be tens of millions of dollars of additional weed control in the United States and Mexico; the benefit would be a "case history" of the consequences of introducing a transgenic fitness-enhancing trait in a crop that readily hybridizes with a local, wild relative.
>
> Are you willing to endorse that experiment?

They were. Even though APHIS-BRS rejected the suggested "experimental approaches" as "flawed," to their credit Upjohn/Asgrow scientists went ahead and did some experiments of the type designed by the academics. Regrettably, they had a difficult time growing the wild plants and the crop-wild hybrids. The sample sizes of the surviving plants were so small that drawing any definitive conclusions was impossible.

In mid-November 1994, I met an APHIS-BRS scientist during the Third International Symposium on the Biosafety Results of Field Tests of Genetically Modified Plants and Microorganisms. We had a lively talk about the squash.

I expressed my concern about the deregulation of ZW-20 being an unfortunate precedent for agbiotech crops, suggesting that APHIS-BRS might want to wait a while to see how well other transgenic crops behaved before releasing one that might drive the evolution of increased weediness in a wild relative. "Norm," he said with a grim grin, "we are taking the large number of public comments seriously. There's *no way* that that squash is going to be deregulated." I left the meeting satisfied that precaution had prevailed. APHIS-BRS deregulated ZW-20 four weeks later. The take-home message? One regulatory scientist does not a policy decision make. When I inquired about that confident statement about the squash to another APHIS-BRS regulator a few years later, I was told curtly, "He doesn't work for us anymore."

Upjohn/Asgrow petitioned APHIS-BRS in 1995 to deregulate a similar variety, CZW-3, engineered for resistance to three viruses: WMV2, ZYMV, and CMV. APHIS-BRS received very few public comments, all positive. The acrimony kindled by the first decision had soured those scientists who had called for more experimental information before allowing environmental release. Those of us who had compromised our teaching and research to provide public comment the first time had learned our lesson.

Now, more than two decades later, we can ask, "How did the grand experiment do?" Obviously, neither of the deregulated virus-resistant crooknecks turned out to be the "Vegetable Marrow That Ate Little Rock." But a few of the interested scientists who had squared off on one side or the other in 1994 took advantage of the fact that they could buy seed of an engineered crop to do the experiments they had discussed. As a result, both the ecology of FLCP pest interactions and the ecological effects of the transgenes are now much better understood. Indeed, the data accumulated since deregulation would have provided a predictive perspective on the real fate of the crop in the last two decades. Let's review the results of that research.

Did long-distance romance enable the transgene to establish itself in FLCP populations? Research using genetic markers has shown that the crop genes are present in free-living FLCP populations, indicating past hybridization. But a group of University of Nebraska scientists surveyed many FLCP populations for years to see if transgene itself has established. They came up empty-handed (Prendeville et al. 2012). Nonetheless, it is hard to say one way or the other whether the transgenic squash has *ever* successfully hybridized with any free-living plant. The transgene might not have ended up in those populations because transgenic varieties have never been as wildly popular as the transgenic varieties of corn, canola, cotton, and soybean. Transgenics make up only 18% of the US summer squash crop (Johnson, Strom, and K. Grillo 2007). The transgene might have occasionally entered natural populations and gone extinct.

Another reason is that more recent research has demonstrated that the conditions under which the transgene would be favored to increase in unmanaged populations are not nearly as common as ecologists and evolutionists anticipated at the time of the ZW-20 deregulation. Some of the initial research was conducted by Hector Quemada, one of the Upjohn/Asgrow scientists who contributed to the petition for deregulation. He assembled a team to conduct a closer examination of the virus infection status of FLCP to ask how seriously virus disease was regulating wild/weedy squash populations. The team surveyed dozens of FLCP populations from Illinois to Texas and chose ten populations to study closely over three consecutive years. To confirm whether a plant was infected with CMV, ZYMV, or WMV2, they used the presence of virus infection symptoms (which, they note, "may not be seen in an infected plant") as well as a more sensitive immunological test strip (similar in concept to the familiar human pregnancy test strip). Every year, they found at least one infected plant in well over half the populations. Infection was often present, but it was never severe. Typically, the majority of plants in a population were not infected by any of the three viruses. Even when they were infected, they were asymptomatic; only 2% of the plants observed displayed viral infection symptoms. Interestingly, other viruses were detected in the free-living populations; CMV, ZYMV, and WMV2 are not the sole viral FLCP pests. The results contradict the conclusion from the initial limited Upjohn/Asgrow survey that viruses are absent from FLCP populations, but they support its other conclusion: that virus disease pressure from ZYMV and WMV2 rarely, if ever, plays a role in determining the fitness of individuals in unmanaged natural populations (Quemada et al. 2008).

In addition to such descriptive studies, other research groups performed various experiments to compare the fitness of FLCP plants, squash plants, and their hybrids—with and without the transgene—in a variety of field environments. The results varied over the experiments. But most studies reported that the transgene had no impact on a plant's fitness, positive or negative, when viruses were absent. When the viruses were present, the conclusions were more complicated than what was reported to APHIS by Wilson and Upjohn/Asgrow. The studies revealed that whether first-generation FLCP × squash hybrids were more fit, less fit, or not significantly different from their wild parents depended strongly on the environment in which they were grown.

Let's focus on a study that simulates natural conditions of FLCP populations that grow outside of squash fields but within pollination distance; that is, plants that were grown without pesticide application and without virus inoculation. Without insecticides, the aphids that can transmit disease-causing viruses and the cucumber beetles that can vector bacterial diseases are free to chow down on experimental plants. Dr. Miruna Sasu's dissertation work at Pennsylvania State University under Professor Andy Stephenson illustrates how ecological genetic interactions can be both surprising and hard to predict a priori. She compared the performance of pure wild plants with transgenic and non-transgenic crop × wild hybrids and plants resulting from backcrosses of the hybrid to wild FLCP. In each year of a three-year experiment, ZYMV naturally swept through the experimental population by aphid transmission. As it did, the heavy infection load led to the typical disease symptoms and slowed growth on plants unprotected by the transgene. Cucumber beetles avoided unhealthy (non-transgenic) plants, feeding mostly on the healthy transgenic plants. Subsequently, the transgenic plants were hammered by the fatal bacterial wilt disease transmitted by the beetles. By measuring and modeling the impacts of all the insect and disease damage, Sasu demonstrated (at least for her study system) that any evolutionary benefits to wild *Cucurbita* conferred by an itinerant virus-resistant transgene introduced by wandering crop pollen would be fully or partly counterbalanced by beetle feeding preference and, consequently, higher levels of a more deadly bacterial disease (Sasu et al. 2009).

The experiments following deregulation have delivered a lot of information unavailable in the mid-'90s. First, they offer the simple reminder that the depth of human ignorance—even for scientists, or *especially* for scientists—is staggering and humbling. Some who opposed deregulation (myself included) were convinced that hybridization and a single adaptive benefit were enough to start the evolutionary ball rolling to increased weediness or invasiveness. We were wrong with regards to this system. Most of us had worked on natural systems and were not cognizant of the large numbers of pest-resistant genes that had been introduced into agronomic crops for years that had the opportunity to enter the populations of wild relatives—apparently without resulting in the evolution of Attila-like hordes. Why should a transgene be any different?

In the interest of documenting ignorance on both sides, some of the "facts" in the petition, its appendices, and the APHIS-BRS evaluations upon which the final determination was made have subsequently been shown to be simply wrong. Here is a single example. It is illuminating to compare the recent thorough multi-year, multi-method Quemada conclusion (Quemada et al. 2008; confirmed and extended by Prendeville et al. 2012) that virus infections are present in most FLCP populations with the APHIS-BRS statement that an initial onetime survey of visual symptoms of a single plant per population was "appropriate, adequate, and **proven** [bold print is that of APHIS] means of determining whether a plant is a significant natural host for a particular virus" (quoted in National Research Council 2002). Also, research has shown that in the absence of the virus and all other pests, that particular transgene is neither evolutionarily beneficial nor detrimental in the systems in which it has been studied (Laughlin et al. 2009). The late twentieth-century assumption (particularly by those who are pro-agbiotech) was that any transgene would always impose a fitness cost in unmanaged environments, proving detrimental in the wild, and be purged by natural selection. Finally, the most significant finding is that while the squash transgene delivers some virus disease resistance to FLCP individuals for certain viruses, the effects of ZW-20's resistance are highly context dependent and will very rarely translate into an unambiguous fitness boost for FLCP. Had those data been present in 1994, I doubt that a single ecologist or evolutionary biologist would have opposed the deregulation of ZW-20.

Years of experiments have taught evolutionary biologists that a single gene change might result in a substantial fitness boost. However, the sciences of plant-pest interactions and invasion biology have now advanced to the point that it is clear that organisms in their homelands are controlled by a suite of pests, not just one. The research of the past two decades revealed that the wild squashes of North America suffer from various viral diseases, sucking aphids, munching beetles, and a fatal bacterial infection. In this case, resistance to a tiny number of disease-causing species was insufficient to tip the evolutionary balance to create a new plant pest. It is unclear whether evolution of resistance to a single pest in an organism's long-term homeland would ever easily convert a benign species into an invasive because the species has had such a long time to accumulate biological enemies over ecological time.

But it's good to be humble and open-minded. The young science of invasive and weedy plant evolution has a long way to go. How problem plants come to be may not follow a single trajectory. Such evolutionary change appears too complex in the case of virus resistance and squash. It was simple in the case of Europe's biggest weed problem, weed beets: a single genetic change in bolting time. However, even that evolutionary pathway would have been hard to anticipate (even by a formal risk assessment) because bolting time itself does not contributing directly to fitness.

Compared to the early days, the current scientific discourse regarding transgenic crop biosafety is now calm, polite, balanced, and better informed. Every major producer of engineered crop seed has its own biosafety scientists. National agbiotech regulatory agencies have added scientists trained in evolution, population genetics, and ecology to their original mix of molecular geneticists, microbiologists, weed scientists, and plant pathologists. Many of the ecologists and evolutionary biologists (like me) interested in biosafety have moved on to related research questions such as the evolution of invasiveness and weediness.

Unplanned sex involving the extraordinary products of agbiotech still deserves attention but depends on the specific products (not the process). The possibility of seed mixing or illicit sex that delivers a Flavr Savr–like gene from one tomato variety to another would hardly evoke a yawn from a biosafety scientist these days. But the concept of a corn plant transformed to produce a pharmaceutical compound in its seeds having an affair with a plant intended for human or animal consumption would yield—and has yielded—concern. Such drug-producing crops are regulated by both APHIS-BRS and the US Food and Drug Administration. About fifteen years ago, corn was the platform of choice for producing pharmaceutical products on the farm (USDA-APHIS 2017). They stimulated a lot of controversy and not a little worry (Fernandez, Crawford. and Hefferan 2002). That's no longer the case because it became clear that it's hard to keep plants' pollen (particularly long-ranging corn pollen) and seeds from sneaking off to where they shouldn't go. The idea of pharmaceutical and other industrial compounds appearing unexpectedly in polenta, corn chips, and cereal was enough to encourage those who are responsible for corn products in the food chain to recommend that biotech companies seek alternate species that won't disperse their pollen so easily. Corn is no longer the go-to choice for producing such chemicals (USDA-APHIS 2017). It turns out to be a good idea because transgenes are being found where they aren't expected, including cases involving corn.

For almost two decades, about once a year, there's a newsworthy case of crop transgenes discovered residing in an unintended location. In some cases, seeds get mixed during plant improvement, in processing facilities, or spilled on the roadside. Other times, transgenic plants engage in unplanned sex. See table 6.2 for some representative examples resulting from these interesting trysts.

Canola (oilseed rape) plants have been the most promiscuous of the bunch. Self-pollinated, wind-pollinated, and insect-pollinated, they also produce smallish (a couple millimeters in diameter) seeds that occasionally escape from a rail car or a truck carrying them to market or a port. Known for its polyunsaturated, heart-healthy vegetable oil, canola/oilseed rape has become one of the top ten world's crops. The combination of its popularity and tiny seeds has resulted in canola becoming an increasingly common roadside weed due to colonization by spilled seeds. Not a particularly nasty weed, canola can be controlled by many herbicides. But the number of effective herbicides is beginning to drop.

TABLE 6.2. *Transgenes on Holiday: Some examples of unplanned sex (VGT) involving genetically engineered plants and unintended mates leading to transgenes out of place**

What? The transgenes and organisms	When? First discovered or reported	Where?
Spontaneous hybridization between canola varieties leads to multiple-herbicide-tolerant (transgenes for glyphosate and glufosinate tolerance) volunteers in agricultural fields.	1998	Various Canadian provinces
Despite Mexico's multi-year moratorium against planting transgenic corn, transgenes (glyphosate tolerance and insect resistance) are detected in traditionally managed landraces, confirming them as descendants of plants that mated with engineered varieties.	2001, with several follow-up studies	Various Mexican states
Spontaneous interspecies hybrids between glyphosate-tolerant canola and the weed wild bird rape.	2003, follow-up study in 2008	Québec, Canada
Spontaneous hybridization between canola varieties leads to multiple-herbicide-tolerant (transgenes for glyphosate and glufosinate tolerance) roadside volunteers. (Note: Japanese buy, but do not grow, transgenic canola—these plants or their ancestors must have spilled from vehicles transporting imported seed from ports.)	2003, with several follow-up studies	Japan
Fruits from non-engineered papaya trees occasionally bear seeds with the transgene for resistance to papaya ringspot virus (due to spontaneous cross-pollination with nearby transgenic trees).	2004	Hawai'i, USA
Spontaneous hybridization between canola varieties leads to multiple-herbicide-tolerant (transgenes for glyphosate and glufosinate tolerance) roadside volunteers.	2004	Manitoba, Canada

(*continued*)

TABLE 6.2. *Continued*

What? The transgenes and organisms	When? First discovered or reported	Where?
Spontaneous hybridization between transgenic creeping bentgrass × weedy creeping bentgrass leads to hybrids and their descendants that bear a not yet deregulated glyphosate-tolerance transgene.	2004, plus several follow-up studies	Oregon, USA
Spontaneous hybridization between transgenic creeping bentgrass × redtop (a weed) leads to hybrids bearing a not yet deregulated transgene for glyphosate tolerance.	2004	Oregon, USA
Spontaneous hybridization between a rice varieties results in a not yet deregulated transgene for glufosinate tolerance found at low frequency in a common, non-engineered rice variety.	2006	Detected in Europe; hybridization occurred in Louisiana, USA
Spontaneous hybridization between canola varieties leads to multi-herbicide-tolerant (transgenes for glyphosate and glufosinate tolerance) roadside volunteers.	2010	North Dakota, USA
Mexico has not formally deregulated genetically engineered cotton, but thousands of acres are grown every year as "field trials." Scientists detect multiple transgenes (glyphosate tolerance, glufosinate tolerance, and two different types of insect resistance) in natural populations of cotton's progenitor, confirming them as descendants of plants that mated with engineered varieties.	2011	Various Mexican states

*Adapted from Ellstrand 2012.

Old-fashioned sex between different volunteer canola varieties has led to the evolution of herbicide tolerance in a similar (but much less dangerous) way that HGT has led bacteria to collect genes for resistance to different antibiotics. Triple-herbicide-tolerant canola plants have been found along the roads of canola-producing regions like Saskatchewan and Alberta in Canada, and North Dakota in the United States, as well as the roads to and from ports of inter-oceanic trade of Canada and Japan (Ellstrand 2012).

While sex has prevailed and transgenes are out and about, the major damage that transgenes on the loose have done is to embarrass the subset of individuals in the agricultural biotechnology industry who have repeatedly assured the public that—with regard to keeping transgenic plants where they should be—everything is under control. For the most part, as reflected in table 6.2, sex has acted to transport transgenes from crop variety to crop variety. In only a very few cases have engineered genes ended up in wild or weed populations. It wasn't a big surprise when glyphosate-tolerant transgenes ended up in interspecies hybrids between canola and the weed bird rape (a close relative to canola) in Québec (Warwick et al. 2008). The official Canadian deregulation documents had predicted it was likely to occur. Those hybrid-derived populations have been monitored for years after farmers stopped growing engineered rapeseed in the region. The transgene's frequency is gradually decreasing.

The most amazing story of transgene escape doesn't involve a food plant, but it is way too exciting to skip. A transgene engineered into the turfgrass species creeping bentgrass made its way out of Oregon field trials and into natural populations of both wild creeping bentgrass and a bentgrass relative called redtop. The field trial plants created by a Scotts Company and Monsanto partnership were engineered for glyphosate tolerance; the idea was to create a "million dollar" turfgrass for putting greens. Golf course personnel could simply apply glyphosate to a putting green and kill every species except the transgenic bentgrass. The phrase "million dollar" refers to the fact that an absolutely uniform, weed-free putting green of pure bentgrass would make the difference between a million-dollar winning putt and one that was deflected by an errant blade of fescue.

In 2003 the bentgrass was proceeding nicely toward deregulation. Scotts was eager to sell it and anticipated little opposition to a non-food transgenic product. So why wait for deregulation to multiply the seed for the market? Scotts contracted farmers in a roughly seventeen-square-mile control area in eastern Oregon and submitted a single field-trial application that was approved by APHIS-BRS. The specific locations of the field trials were kept confidential. While the reason is not clear, one theory is that it was to discourage radical protestors dressed like Franken-grass from disrupting production. The farmers went to work to ramp up material to sell when deregulation occurred.

Creeping bentgrass is wind-pollinated. We already know that wind-dispersed pollen can, depending on the species, go a long way to fertilize another compatible plant. Wild relatives of creeping bentgrass, including wild types of the very same species, are part of the natural flora of eastern Oregon. Without some isolation efforts, pollen would serve as a vehicle to move the regulated transgene into a cross-compatible wild population via sex. As part of the field-test application, Scotts and Monsanto would be obligated to describe how the transgene would be contained. That information isn't readily available to the public, but we can make some educated guesses. Here's what was known in the literature. A couple of years before this extensive field release of the transgenic bentgrass, a couple of scientists at Pure-Seed Testing Inc. of Hubbard, Oregon, reported a set of experiments to estimate how successful bentgrass hybridization rates change over distance (Wipff and Fricker 2001). They used a transgenic variety tolerant to a different herbicide, glufosinate, for their marker—a few hundred plants planted as two seed nurseries to serve as the center of their plots, serving as fathers. They planted non-transgenic plants, serving as mothers, at varying distances from the nurseries. They harvested seeds from the mothers in each of two different years. The resulting seedlings were planted and screened with three applications of glufosinate herbicide, roughly four million seedlings in all. They modeled their data and found that transgenic pollen hybridization rates varied, but at the maximum, hybridization rates should drop to the 0.02% level at 4,296 feet (about 0.8 of a mile).

The EPA's Oregon-based Dr. Lidia Watrud caught wind (so to speak) of this massive field trial. She and her team knew that the maximum estimated cross-pollination distance of creeping bentgrass was large relative to the size and shape of the control area. They had a hunch that the isolation distance of even 0.8 miles was insufficient to contain the pollen. Watrud's team collected seeds from natural and experimentally planted wild plants at various distances from the edge of the control area. They germinated the seeds and first tested the seedlings for glyphosate resistance with two cycles of spraying. Those surviving a glyphosate application were retested for the presence of the transgene, which would confirm they were fathered by the engineered putting-green strain. All of the seedlings that survived contact with glyphosate had the transgene. Because the specific paternal population sites were unknown, Watrud's group made conservative estimates of the maximum distance that transgenic pollen had traveled to father the hybrid seed: thirteen miles for spontaneous crosses to wild bentgrass and eight and a half miles for redtop (Reichman et al. 2006). Because the field sites were not at the very edge of the control area, those measurements are underestimates. At the time of Watrud's publication (Watrud et al. 2004), the report of thirteen miles was a world record, the *longest recorded interplant mating distance in scientific history*. Not bad for an organism that has no feet, wings, or fins!

The Scotts Company has been fined and has made efforts to eradicate free-living transgenic plants. Nonetheless, they persist. The frequency of the transgene in wild populations has been measured periodically both within and beyond the former control area. At present, Carol Mallory-Smith's laboratory at Oregon State University has taken the lead in studying the unfolding transgenic bentgrass story. In contrast to the canola–bird rape hybrids discussed above, the transgene appears to be slowly increasing in frequency, even though the wild populations occur in locations where glyphosate is rarely used (Reichman et al. 2006; Zapiola et al. 2008). (For much more on transgenic turfgrass follies, see Snow [2012].)

In 2015 Scotts and Monsanto withdrew their 2003 petitions and submitted a new one for deregulation.[5] In September of that year, APHIS-BRS, Scotts, and the USDA reached a Memorandum of Understanding (MOU) and a Memorandum of Agreement (MOU) that Scotts would act

5 www.aphis.usda.gov/brs/aphisdocs/15_30001p.pdf.

as a resource for helping to mitigate the problems of farmers and irrigation districts burdened by the new weed through at least 2018. At that time both Scotts and Monsanto agreed to neither propagate nor commercialize the plants in the future. In January 2017, APHIS-BRS approved the Scotts and Monsanto petition for deregulation of the herbicide-tolerant bentgrass.[6] But this approval came with the recognition that the MOU and MOA of 2015 would still stand.

Natural transgenic creeping bentgrass bearing a gene for glyphosate resistance is hardly a Franken-grass, but it is one more species that can no longer be controlled with glyphosate. And because it cannot be controlled by glyphosate, it is beginning to choke irrigation canals in Oregon and elsewhere. Glyphosate is one of very few herbicides permitted to be used in such situations (Zapiola and Mallory-Smith 2017). Although the story doesn't involve a food crop (the seeds of a closely related species are known to be edible), this example of involving a domesticated turf crop illustrates how far crop plants will go to find a mate. Long-distance romance, indeed!

Sex happens, regardless of whether a plant is transgenic or not. There are dozens of examples of sex delivering non-engineered crop genes into plants for which they were not intended. If you have ever bought a packet of white radish seeds and found a few plants with red or purple roots, you'll see the effects of unintended sex. These crop-to-crop trysts have long been considered a fact of life for plant breeders, who try to minimize what they call "contamination."

6 See https://www.aphis.usda.gov/aphis/newsroom/stakeholder-info/sa_by_date/sa-2017/sa-01/sa_cbg_rod and links therein.

Furthermore, when most of the world's crops come within pollinating distance of wild relatives, at least a few illicit cross-pollinations occasionally occur, both into the crop plants and into the wild ones (Ellstrand 2003). For traditional breeders, gene flow from the crop into wild relatives was of no consequence. Until the evolution of Europe's weed beets, the rare gene flow from wild relatives into the crop was usually considered another type of genetic contamination to minimize as an issue of varietal purity.

But transgenes can be patented and pilfered. The companies that create and market them are uneasy about their products ending up in the wrong hands. Various supplementary genetic-engineering strategies have been proposed and studied to keep these products of intellectual property down on the farm, both in terms of securing property from theft as well as putting an end to embarrassing news reports regarding transgenics "sowing their wild oats," so to speak. These strategies include transgenes for seed sterility (the so-called "terminator" gene), asexuality, and male sterility, as well as methods to place the transgene in the chloroplast in those plants that have strict maternal inheritance. None have yet proven effective enough or otherwise worthwhile to move toward deregulation (National Research Council 2004). Until a solution is found to contain transgenes effectively, sex not only happens, but it rules.

The nature of genetic engineering and transgenic crops moves ahead by fits and starts. The major crops, important companies, primary traits, and significantly producing countries involved varied little from the late 1990s to the early 2010s. In the years that I wrote this book, things started getting dynamic. It is hard to say if that trend will continue. The information in this chapter is accurate to mid-2017. The only thing we can be sure of is that the importance of sex will not disappear from the food we eat. The epilogue takes a look to the future of the interaction of plant sex and our food.

Transgenic—or Not—Tacos

Some folks are going out of their way to avoid eating genetically engineered foods for a variety of reasons. For those with that bent, resources abound, for example, *The Non-GMO Cookbook* (Pineau and Westgate 2013). But what about folks on the other side? I've tried in vain to find *The GMO Cookbook* or its equivalent. Just for fun, here is a pair of recipes: one that minimizes transgenic products, and one that maximizes them. At the moment, avoiding transgenic foods is easy. Buy organic. Buy non-GMO-labeled food.

Or more easily? Want to radically reduce transgenic consumption, but don't want to be fussy and don't care if you go exactly to 0%? Simply remove processed foods from your life.

In contrast, it is not at all easy to create a recipe of entirely transgenic foods. There just are too few engineered species on the market. I searched among my favorite recipes to see what I could tweak to come up with a suitable set of contrasting ingredients.

Pamela Ronald and Raoul Adamchak (2008), geneticist and organic farmer, respectively, coauthored *Tomorrow's Table: Organic Farming, Genetics, and the Future of Food.* The book is a straightforward, side-by-side examination of genetic engineering and organic farming. It is one argument for mindful use of all of the tools we have to meet the crisis of feeding billions of souls in an environmentally sustainable way. The following couplet of recipes are inspired by that book's "Pam & Raoul's Tofu Tortillas," one of my favorites.

I was able to accomplish the transgenic-free recipe but could not create an all-transgenic recipe, settling instead for a transgenic-enhanced alternative. Here they are, with commentary on the ingredients (accurate as of late 2016).

Transgenic Enhanced Tacos (accurate as of late 2016), optimized in favor of transgenic products, with a bit of cheating (see comments on onion). Do not use organic or non-GMO products, which are always highly non-transgenic.

Ingredients	Commentary
Canola oil	Transgenic canola (both *Brassica rapa* and *B. napus*) is commercially grown in three countries. In the United States and Canada, transgenics account for more than 95% of global production.
12 corn tortillas	Corn (aka maize) is one of the three global transgenic mega-crops. Transgenic varieties are commercially grown in seventeen countries on five continents. In the United States, Brazil, Argentina, and Canada, they make up the vast majority of corn produced. The majority of the corn produced in the world isn't used for direct human consumption but for animal feed, biofuel, processed foods, and industrial products.
2 cups grated hard cheese	Originally, most hard cheeses were created with rennin (the enzyme chymosin, a coagulating compound originally obtained from the stomachs of slaughtered newly born calves). Nowadays, in developed countries such as the United States, Canada, and even much of Europe, chymosin is produced by bacteria engineered with the bovine gene does the job. Alternatives to baby cattle and genetic engineering are vegetable rennet from plants and microbial rennet from fungi. The cheeses based on genetically engineered bacteria predominate in the markets of North America and Europe.

Ingredients	Commentary
1 lb. grated firm tofu	Soybean is one of another of the three global transgenic mega-crops. (The third is cotton.) Tofu is soybean curd. Transgenic soybean is commercially grown in eleven countries. In the United States, Brazil, Argentina, Canada, and Paraguay, it makes up the vast majority of soy produced. Even though tofu is an important food in much of East Asia, at the time of this writing not a single Asian country has produced transgenic soy. The majority of the soy produced in the world isn't used for direct human consumption but separated into meal, almost exclusively used for animal feed and oil, which is used for biodiesel and industrial products, or makes its way to human consumption indirectly in processed foods or as one of several vegetable oils.
3 tablespoons soy sauce	Traditional soy sauce is made from water, wheat, soy, and salt. Water and salt are obviously non-transgenic. Neither is commercially available wheat.
2 tablespoons minced fresh onion	I'm cheating here. For superior flavor, these tacos require a savory vegetable in the onion family. The only species in that family that has been even field-tested is onion itself. Its last field tests were years ago in the United States and New Zealand. You will not be able to obtain transgenic onion.
4 tablespoons fresh papaya seeds (from fresh fruit)	Transgenic papaya is grown for consumption in China and the United States, specifically Hawaiʻi. To be sure that your fruit is transgenic, buy one of the following varieties: Rainbow, SunUp, or Sunrise.
¼ cup dry roasted edamame	Edamame are immature soybean fruits. The seeds from the fruits can be bought in the dry, roasted form. See full commentary on soybean in the tofu entry above.

Fry the onion in a pan until transparent. Add tofu, edamame, and soy sauce. Stir fry. When brown, keep warm. Fry tortillas on both sides in the oil. Sprinkle with the grated cheese. Fill tortillas with fried tofu mixture. Sprinkle on papaya seeds (note, they are spicy). Serve with corn relish.

Transgenic-Free Tacos (accurate as of late 2016)

All of the following ingredients should be transgene free.

Ingredients	Commentary
Olive oil	Transgenic olive grows in field tests in Europe. There's no apparent interest in moving toward deregulation and commercialization.
6 wheat tortillas	Many types of transgenic wheat have undergone hundreds of field trials in the United States and elsewhere. In 2004 wheat genetically engineered for glyphosate resistance was approved for human consumption in Australia, Colombia, the United States, and Canada. However, it has never been deregulated for agricultural production. Therefore, it cannot have been commercialized.
1 cup acid-precipitated cheese transgenic (e.g., queso blanco)	Acid-precipitated soft cheeses never involve rennet (see above) in their preparation.
2 garlic cloves, minced	Garlic has been transformed in the lab and grown in the greenhouse, but apparently not yet in the field.
½ teaspoon chili pepper flakes	Transgenic sweet peppers have been approved to be grown in China for food, but it's not clear whether they were ever commercialized. I could not find any record of transgenic chili pepper (the same species) field tests.
1 lb. shredded cooked chicken	Many kinds of transgenic animals have been created and studied under lab conditions. No transgenic bird or mammal has been approved for human consumption, and there's no indication of any in the pipeline.
¼ cup sunflower seeds	The last transgenic sunflower field trial was conducted more than a decade ago. Thus, no product is moving toward deregulation and commercialization.
1 tablespoon cumin	"Proof-of-concept" transgenic cumin has been created in the lab, but it probably will be years before a transgenic type will be created for field-testing and longer still for one worthy of commercialization.
1 tablespoon beef broth	See commentary for chicken above.

Ingredients	Commentary
1 tablespoon balsamic vinegar	Balsamic vinegar is a grape product from Italy. Various types of transgenic grapes have been field-tested in the United States and elsewhere including Italy. Transgenic grapes seem far from deregulation anywhere, and especially so in Europe.
1 tablespoon molasses from sorghum	Various types of transgenic sorghum have been field-tested in the United States and Africa. Nothing appears close to deregulation. Molasses from sugarcane was intentionally avoided. There are a lot more engineered sugarcane trials going on at the moment, on just about every tropical continent.

Fry the garlic in a pan until transparent. Add chicken, chili flakes, cumin, broth, vinegar, and molasses. Stir fry. When brown, keep warm. Fry tortillas on both sides in the oil. Spread with cheese. Fill tortillas with fried chicken mixture. Salt to taste. Serve with avocado, sugar-free and oil-free salsa, spring green mix, chopped tomatoes (none of those have transgenic products in the marketplace as I write these words).

Epilogue

BACK IN THE GARDEN

What you put in your mouth is connected to the rest of the world.
—Source unknown

Soda crackers rarely make a good meal. Lists of facts to be remembered for the exam rarely make a good read. Likewise, cotton candy may be tasty and exciting, but, in the end, it's rarely satisfying. The hype *du jour* also leaves that empty feeling. Building a meaningful romance with food means more than counting calories or the passions of panic and pleasure. It is impossible to build a romance based on absolutes.

A lasting romance comes from understanding. Understanding is more than knowledge; it is the substrate for the growth of wisdom. With regards to what you put in your mouth, it starts with the understanding that the food is connected to the rest of the world. Our spin of the kaleidoscope has focused a single sliver of that connection, the role of sex in getting food to your table.

One of the goals of *Sex on the Kitchen Table* is to reveal that we actually never left the Garden. Now you are ready to enjoy the Garden on your own, relishing in the interrelatedness of romance and food, sex and sustenance. It's a lot of fun.

As you know, science is neither scary nor the sole property of the nerdy. You've got the tools now to understand where your food comes from. You can blend what you learn with your values to determine your own stand on some of today's food-related controversies.

Understanding how nature works is one definition of science. Technology and science are not the same. Technology is a set of tools derived from scientific information and the application of those tools. Like science, "technology accomplishes nothing by itself. It has no will or moral purpose, any more than the law of gravity does" (Charles 2005).

Let's consider genetic engineering and the newly minted gene-editing techniques. As I like to tell my students, "Genetic engineering is a tool. So is a hammer. A hammer can be used nobly—for example, to build a home for the homeless. A hammer can have an ominous use—for example, bashing in a professor's brain. I am not 'for' or 'against' a hammer. But I do have some opinions about its use!" The same is true for the technology of plant genetic engineering.

At the moment the world faces a daunting set of challenges. The combination of too many people and too much consumption has created an unsustainable situation (Brown 2009). I care about the future of the students in my classes. We need to be mindful of every tool in our tool kit; nothing should be excluded from consideration. With regards to feeding folks, every option needs to be considered from family planning, especially the families whose individuals have the greatest impact on the Earth's resources, to reduced consumption of inefficiently created calories from warm-blooded animals to skillful deployment of sustainable agriculture techniques. Included in the list is getting "two blades of grass to grow where one grew before" (Swift [1726] 1999)

FIGURE 7.1 Back in the Garden.

via the most appropriate method of plant improvement, whether the domestication of new food plants, traditional breeding, or the new fields of synthetic biology, molecular breeding, and genetic engineering. Rather than addressing future crises fearfully, we can often do it joyfully. As a friend of mine likes to say, "Save the world while having fun."

Here's one example of a fun food that is one of many potential wedges against hunger. Some of the most tasty and satisfying foods I've eaten are insects. Deep-fried beetle grubs in Kunming, Yunnan, China. Butter and garlic sautéed *escamoles* in Morelia, Michoacán, Mexico. Chocolate-dipped crickets in Riverside, California, USA. This is good stuff (Deroy 2015)! It also happens to be good for the future of the planet. Cold-blooded insects can produce a pound of high-quality animal protein with a tiny fraction of the water or energy consumed to produce a pound of beef. They can feed a lot more folks with a lot fewer resources. No quarrel against beef here, particularly the sustainably raised type.

A Garden-based approach is a mindful one. So when we pick up a tool, we consider it before putting it to use. We feel the hammer's heft. We hold it for a moment and set our intent on how to use it properly. Maybe it is the perfect tool and we deploy it. A joyful arc and the satisfying plunge of the nail. Perhaps we recall a better tool in the kit, and we set the hammer down. In short, the hammer of the technologies that comprise plant genetic engineering is here to stay in our tool kit. It has already proven a good solution for certain problems and has already failed for others. It must be part of the solution, but—just like any other technology—we must make sure that we mindfully decide when to use it and to mindfully control how we use it, so that we avoid hitting our finger while going for the nail.

Acknowledgments

It has taken more than a village to create this book. You wouldn't be reading these words if not for the generous community of varied support I received prior to and during their creation. I have already described the encouragement I received from my parents and early scientific mentors, David Nanney, Don Levin, and Janis Antonovics. Many more have taught me about plants, sex, genetics, and food. These include dozens of colleagues at University of California, Riverside (UCR), both senior and junior. Closest to home, members of my lab family, my friends who named themselves "Team Ellstrand," have been sources of lifelong support and inspiration. Janet Clegg, Paul Arriola, Detlef Bartsch, Lesley Blancas, Jutta Burger, Lisa Ciano-Gandy, Bernie Devlin, Diane Elam, Michelle Gadus, Karen Goodell, Roberto Guadagnuolo, Subray Hegde, Joanne Heraty, Sylvia Heredia, Timothy Holtsford, Terrie Klinger, Janet Leak-Garcia, Jennifer Lyman, Diane Marshall, Marlyce Myers, John Nason, Maile Neel, Deborah Pagliaccia, John Peloquin, Robert Podolsky, Caroline Ridley, Jeffrey Ross-Ibarra, Toni Siebert, Katia Silvera, Shana Welles, Li Yao, and Melinda Zaragosa. It's an honor to have worked with them. Dozens of UCR faculty and staff have mentored me, too many to mention. The list continues to grow with the new crop of young scholars in ecology and evolutionary biology. They are smart, positive, and energetic, setting a perfect example for those of us who have been at UCR awhile. Outside of UCR, sabbatical hosts have been generous with their time and resources. The list includes Michael Arnold, Francisco Ayala, Herbert Baker, David Lloyd, Honor Prentice, Loren Rieseberg, Jeffrey Ross-Ibarra, Shelly Schuster, and Monte Slatkin. My Chinese "younger twin brother," Baorong Lu, deserves special mention for his friendship and scientific insights; I have cherished our ongoing collaboration.

Over my career, financial support for my food plant/sex research came from a stone soup of sources: UCR's Agricultural Experiment Station, a Fulbright Fellowship to Sweden, a John Simon Guggenheim Memorial Fellowship, as well as grants, contracts, and gifts from NSF (particularly, NSF OPUS grant DEB-1020799), the USDA, UC MEXUS, the California Avocado Commission, the Swedish Forestry and Agricultural Research Council, the University of California's Division of Agricultural and Natural Resources, the University of California Biotechnology Research and Education Program, the University of California Germplasm Resource Conservation Program, the California Cherimoya Association, and the California Rare Fruit Growers.

I am indebted to a team of experts who sacrificed their busy time to critique my core chapters: Cynthia Jones (chapter 2), Francis Xavier Asiimwe (chapter 3), Lauren Garner (chapter 4), Detlef Bartsch (chapter 5—beets), Subray Hegde (chapter 5—crop improvement), Peggy Lemaux (chapter 6—agricultural biotechnology), and Andy Stephenson (chapter 6—squash). Lots of bits and pieces of help came from other folks: Gary Bergstrom, Pierre Boudry, Sandra Knapp,

Amy Litt, Carol Lovatt, Deborah Pagliaccia, Diana Pilson, Hector Quemada, Allison Snow, and Larry Venable. But I'll take responsibility for the errors that may remain.

A good fraction of this book was written at "my office away from my office." Thanks to Darren Conkerite for creating and maintaining "Riverside's Living Room," the Back to the Grind Coffeehouse, to be a quiet escape for creative endeavors.

Some special thanks to the following: My editor at University of Chicago Press, Christie Henry, whose enthusiasm and optimism added the necessary energy to keep the project going, a rare gem in the world of scientific publishing; novelist and professor Susan Straight, who—after comforting her dog through a stormy night—still had the heart and energy to give me a pep talk to guide me through a particularly rough time in the genesis of the pages you hold; University of Chicago Press Senior Manuscript Editor Erin DeWitt improved the manuscript's clarity; Dr. Sylvia Heredia, who was willing to share her talent for scientific illustration; artist Beverly Ellstrand for her contribution to the frontispiece art; and most importantly, those who believe in me and put up with me on a regular basis: my favorite colleague and spouse, Tracy Kahn, as well as my pop culture consultant and son, Nathan Ellstrand.

To all the folks named above and the many I fear I may have forgotten, I cannot thank you all enough.

Literature Cited

Acuña, R., B. E. Padilla, C. P. Flórez-Ramos, J. D. Rubio, J. C. Herrera, P. Benavides, S.-J. Lee et al. 2012. Adaptive horizontal transfer of a bacterial gene to an invasive insect pest of coffee. *Proceedings of the National Academy of Sciences USA* 109:4197–202.

Alston, J. M., and G. P. Pardey. 2014. Agriculture in the global economy. *Journal of Economic Perspectives* 26:121–46.

Andersson, M. S., and M. C. de Vicente. 2010. *Gene flow between crops and their wild relatives.* Baltimore, MD: Johns Hopkins University Press.

Antonovics, J., and N. C. Ellstrand. 1985. The fitness of dispersed progeny: Experimental studies with *Anthoxanthum.* In *Genetic differentiation and dispersal in plants*, ed. P. Jacquard, J. Heim, and J. Antonovics, pp. 369–81. Berlin, Germany: Springer-Verlag.

Avery, O. T., C. M. MacLeod, and M. McCarty. 1944. Studies on the chemical nature of the substance inducing transformation of pneumococcal types. *Journal of Experimental Medicine* 79:137–58.

Bakker, H. 1999. *Sugar cane cultivation and management.* New York: Kluwer Academic.

Barnosky, A. D., N. Matzke, S. Tomiya, G. O. U. Wogan, B. Swartz, T. B. Quental, C. Marshall et al. 2011. Has the Earth's sixth mass extinction already arrived? *Nature* 471:51–57

Barraclough, T. G., C. W. Birky, and A. Burt. 2003. Diversification in sexual and asexual organisms. *Evolution* 57:2166–72.

Barron, C. 2016. Belgian man's pumpkin sets world record at a whopping 2,624 pounds. *Washington Post*, October 16.

Beachy, R. N., S. Loesch-Fries, and N. E. Tumer. 1990. Coat protein-mediated resistance against virus infection. *Annual Review of Phytopathology* 28:451–72.

Bennett, K. D. 2013. Is the number of species on earth increasing or decreasing? Time, chaos and the origin of species. *Paleontology* 56:305–25.

Bergh, B. O. 1968. Cross-pollination increases avocado set. *California Citrograph* 53(3):97–100.

———. 1973. The remarkable avocado flower. *California Avocado Society 1973 Yearbook* 57:40–41.

———. 1992.The origin, nature, and genetic improvement of the avocado. *California Avocado Society 1992 Yearbook* 76:61–75.

Bergh, B. O., and C. D. Gustafson. 1958. Fuerte fruit set as influenced by cross-pollination. *California Avocado Society 1958 Yearbook* 42:64–66.

———. 1966. The effect of adjacent trees of other avocado varieties on Fuerte fruit set. *Proceedings of the American Society for Horticultural Science* 89:167–74.

Biancardi, E., L. G. Campbell, G. N. Skaracis, and M. de Biaggi. 2005. *Genetics and breeding of sugar beet.* Enfield, NH: Science Publishers.

Biancardi, E., L. W. Panella, and R. T. Lewellen. 2012. Beta maritima. *The origin of beets.* New York: Springer.

Bijlsma, R., R. W. Allard, and A. L. Kahler. 1986. Non-random mating in an open-pollinated maize population. *Genetics* 112:669–80.

Blackburn, F. 1984. *Sugar-cane.* Harlow, UK: Longman.

Boudry, P., K. Broomberg, P. Saumitou-Laprade, M. Mörchen, J. Cuguen, and H. Van Dijk. 1994. Gene escape in transgenic sugar beet: what can be learned from molecular studies of weed beet populations? In *Proceedings of the 3rd international symposium on the biosafety results of field tests of genetically modified plants and microorganisms*, ed. D. D. Jones, 75–87. Oakland: University of California Division of Agriculture and Natural Resources.

Boudry, P., M. Mörchen, P. Saumitou-Laprade, P. Vernet, and H. Van Dijk. 1993. The origin and evolution of weed beets: Consequences for the breeding and release of herbicide-resistant transgenic sugar beets. *Theoretical and Applied Genetics* 87:471–78.

Bourdain, A. 2000. *Kitchen confidential: Adventures in the culinary underbelly.* New York: Bloomsbury.

Brown, L. 2009. *Plan B 4.0 Mobilizing to save civilization.* New York: Norton.

Can-Alonso, C., J. J. G. Quezada-Euán, P. Xiu-Ancona, H. Moo-Valle, G. R. Valdovinos-Nuñez, and S. Medina-Peralta. 2005. Pollination of "criollo" avocados (*Persea americana*) and the behavior of associated bees in subtropical Mexico. *Journal of Apicultural Research* 44:3–8.

Carman, H. F., and R. J. Sexton. 2007. The 2007 freeze: Tallying the toll two months later. *Agriculture and Resource Economics Update* 10(4):5–8.

Carroll, L. 1871. *Through the looking-glass and what Alice found there.* London: Macmillan.

Chanderbali, A. S., D. E. Soltis, P. E. Soltis, and B. N. Wolstenhome. 2013. Taxonomy and botany. In *The avocado: Botany, production, and uses*, 2nd ed., ed. B. Schaffer, B. N. Wolstenholme, and A. W. Whiley, pp. 31–50. Wallingford, UK: CABI Publishing.

Chandler, W. H. 1958. *Evergreen orchards.* Philadelphia: Lea and Febiger.

Chapman, P. 2007. *Bananas: How the United Fruit Company shaped the world.* Edinburgh: Canongate Books.

Charles, D. 2001. *Lords of the harvest: Biotech, big money, and the future of food.* Cambridge, MA: Perseus.

———. 2005. *Master mind: The rise and fall of Fritz Haber, the Nobel laureate who launched the age of chemical warfare.* New York: Ecco.

Coffey, M. D. 1987. *Phytophthora* root rot of avocado. *Plant Disease* 71:1046–52.

Cokinos, C. 2000. *Hope is the thing with feathers.* New York: Penguin Putnam.

Cox, P. A. 1988. Hydrophilous pollination. *Annual Review of Ecology and Systematics* 19:261–80.

Crepet, W. L., and K. J. Niklas. 2009. Darwin's second "abominable mystery": Why are there so many angiosperm species? *American Journal of Botany* 96:366–81.

Darwin, C. R. 1868. *The variation of animals and plants under domestication.* London: John Murray.

———. 1876a. *The different forms of flowers on plants of the same species.* London: John Murray.

———. 1876b. *The effects of cross and self fertilisation in the vegetable kingdom.* London: John Murray.

———. 1885. *The various contrivances by which orchids are fertilized by insects.* New York: Appleton.

———. (1859) 1902. *On the origin of species by means of natural selection.* London: John Murray.

Davenport, T. L., P. Parnitzki, S. Fricke, and M. S. Hughes. 1994. Evidence and significance of self-pollination of avocados in Florida. *Journal of the American Society of Horticultural Science* 119: 1200–207.

Degani, C., R. El-Batsri, and S. Gazit. 1997. Outcrossing rate, yield, and selective fruit abscission in "Ettinger" and "Ardith" avocado plots. *Journal of the American Society of Horticultural Science* 122:813–17.

Degani, C., A. Goldring, and S. Gazit, 1989. Pollen parent effect on outcrossing rate in "Hass" and "Fuerte" avocado plots during fruit development. *Journal of the American Society of Horticultural Science* 114:106–11.

Degani, C., A. Goldring, S. Gazit, and U. Lavi, 1986. Genetic selection during abscission of avocado fruitlets. *HortScience* 21:1187–88.

Delfelice, M. S. 2003. The black nightshades: *Solanum nigrum* L. *et al.*—Poison, poultice, and pie. *Weed Technology* 17:421–27.

Deroy, O. 2015. Eat insects for fun, not to help the environment. *Nature* 521:395.

Desplanque, B., P. Boudry, K. Broomberg, P. Saumitou-Laprade, J. Cuguen, and H. Van Dijk. 1999. Genetic diversity and gene flow between wild, cultivated and weedy forms of *Beta vulgaris* L. (Chenopodiaceae), assessed by RFLP and microsatellite markers. *Theoretical and Applied Genetics.* 98:1194–201.

Draycott, A. P. 2006. Introduction to *Sugar beet*, ed. A. P. Draycott, pp. 1–8. Oxford: Blackwell.

Dreher, M. L., and A. J. Davenport. 2013. Hass avocado composition and potential health effects. *Critical Reviews in Food Science* 53:738–50.

Ellstrand, N. C. 1992. Sex and the single variety. *California Grower* 16(1):22–23.

Ellstrand, N. C. 2003. *Dangerous liaisons? When cultivated plants mate with their wild relatives.* Baltimore, MD: Johns Hopkins University Press.

Ellstrand, N. C. 2012. Over a decade of crop transgenes out-of-place. In *Regulation of Agricultural Biotechnology: The United States and Canada*, ed. C. A Wozniak and A. McHughen, pp. 123–35. Dordrecht: Springer.

Ellstrand, N. C., and J. Antonovics. 1985. Experimental studies on the evolutionary significance of sexual reproduction. II. A test of the density-dependent selection hypothesis. *Evolution* 39:657–66.

Ellstrand, N. C., and K. W. Foster. 1983. Impact of population structure on the apparent outcrossing rate of grain sorghum (*Sorghum bicolor*). *Theoretical and Applied Genetics* 66:323–27.

Ellstrand, N. C, S. M. Heredia, J. A. Leak-Garcia, J. M. Heraty, J. C. Burger, L. Yao, S. Nohzadeh-Malakshah et al. 2010. Crops gone wild: Evolution of weeds and invasives from domesticated ancestors. *Evolutionary Applications* 3:494–504.

Ellstrand, N. C., and C. A. Hoffman. 1990. Hybridization as an avenue of escape of engineered genes. *BioScience* 40:438–42.

Ernst, A. A., A. W. Whiley, and G. S. Bender. 2013. Propagation. In *The avocado: Botany, production, and uses*, 2nd ed., ed. B. Schaffer, B. N. Wolstenholme, and A. W. Whiley, pp. 243–67. Wallingford, UK: CABI Publishing.

Feldman, M., F. G. H. Lupton, and T. E. Miller. 1995. Wheats. In *Evolution of crop plants*, 2nd ed., ed. J. Smartt and N. W. Simmonds, pp. 184–92. Harlow, UK: Longman.

Fernandez, M., L. Crawford. and C. Hefferan. 2002. *Pharming the field: A look at the benefits and risks of bioengineering plants to produce pharmaceuticals.* Philadelphia: Pew Charitable Trust.

File, A. L., G. P. Murphy, and S. A. Dudley. 2011. Fitness consequences of plants growing with siblings: Reconciling kin selection, niche partitioning and competitive ability. *Proceedings of the Royal Society B* 279: 209–18.

Fink, G. R. 2005. A transforming principle. *Cell* 120:153–54.

Flores, D. 2015. *Mexican avocado industry continues to enjoy strong growth.* US Department of Agriculture (USDA). Foreign Agricultural Service (FAS). GAIN Report No. MX5050.

Flot, J.-F., B. Hespeels, X. Li, B. Noel, I. Arkhipova, E. G. J. Danchin, A. Hejnol et al. 2013. Genomic evidence for ameiotic evolution in the bdelloid rotifer *Adineta vaga. Nature* 500:453–57.

Forbush, E. H. 1936. Passenger pigeon. In *Birds of America*, ed. T. G. Pearson, pp. 39–46. Garden City, NY: Garden City Books.

Ford-Lloyd, B. 1995. Sugarbeet and other cultivated beets. In *Evolution of crop plants*, 2nd ed., ed. J. Smartt and N. W. Simmonds, pp. 35–40. Harlow, UK: Longman.

Francis, S. A. 2006. Development of sugar beet. In *Sugar beet*, ed. A. P. Draycott, pp. 9–29. Oxford: Blackwell.

Frank, D. 2005. Bananeras: Women transforming the banana unions of Latin America. Cambridge: South End Press.

Frundt, H. J. 2009. *Fair bananas: Farmers, workers, and consumers strive to change an industry.* Tucson: University of Arizona Press.

Garner, L. C., V. E. T. M. Ashworth, M. T. Clegg, and C. J. Lovatt. 2008. The impact of outcrossing on yields of "Hass" avocado. *Journal of the American Society of Horticultural Science* 133:648–52.

Griffith, F. 1928. The significance of pneumococcal types. *Journal of Hygiene* 27: 113–59.

Group of Reproductive Development and Apomixis. 1998. "Bellagio Declaration." Laboratorio Nacional de Genomica Para la Biodiversidad. http://langebio.cinvestav.mx/?pag=424.

Hand, M. L., and A. M. G. Koltunow. 2014. The genetic control of apomixis: Asexual seed formation. *Genetics* 19. 7:441–50.

Harper, J. L. 1977. *Population biology of plants.* London: Academic Press.

Heywood, V. H., R. K. Brummitt, A. Culham, and O. Seberg. 2007. *Flowering plant families of the world.* Rev. ed. Richmond Hill, ON: Firefly Press.

Hodgson, R. W. 1930. The California avocado industry. *California Agriculture Extension Series Circular* #43.

———. 1947. The California avocado industry. *California Avocado Society 1947 Yearbook* 32:35–39.

Hokanson, K. E., N. C. Ellstrand, A. G. O. Dixon, H. P. Kulembeka,

K. M. Olsen, and A. Raybould. 2016. Risk assessment of gene flow from genetically engineered virus resistant cassava to wild relatives in Africa: An expert panel report. *Transgenic Research* 25: 71–81.

Holm, L. G., D. L. Plucknett, J. V. Pancho, and J. P. Herberger. 1977. *The world's worst weeds: Distribution and biology*. Honolulu: University Press of Hawaii.

Ish-Am, G., and D. Eiskowich. 1993. The behaviour of honey bees (*Apis mellifera*) visiting avocado (*Persea americana*) flowers and their contribution to its pollination. *Journal of Apicultural Research* 32:175–86.

———. 1998. Low attractiveness of avocado (*Persea americana* Mill.) flowers to honeybees (*Apis mellifera* L.) limits fruit set in Israel. *Journal of Horticultural Science and Biotechnology* 73:195–204.

James, C. 2015. *Global status of commercialized biotech/GM crops: 2015*. ISAAA Brief No. 51. Ithaca, NY: ISAAA.

Jefferson, R. A. 1994. Apomixis: A social revolution for agriculture? *Biotechnology and Development Monitor* 19:14–16.

Johnson, S., S. Strom, and K. Grillo. 2007. *Quantification of the impacts on US agriculture of biotechnology-derived crops planted in 2006*. Washington, DC: National Center for Food and Agricultural Policy.

Jones, J. B. 2008. *Tomato plant culture*. 2nd ed. Boca Raton, FL: CRC Press.

Kahn, T. L., and D. A. DeMason. 1986. A quantitative and structural comparison of *Citrus* pollen tube development in cross-compatible and self-incompatible gynoecia. *Canadian Journal of Botany* 64:2548–55.

Karttunen, F. E. 1992. *An analytical dictionary of Nahuatl*. Rev. ed. Norman: University of Oklahoma Press.

Kaul, M. L. H. 2012. *Male sterility in higher plants*. Berlin: Springer-Verlag.

Keeling, P. J., and J. D. Palmer. 2008. Horizontal gene transfer in eukaryotic evolution. *Nature Reviews Genetics* 9:605–18.

Kelly, A. F., and R. A. T. George. 1998. *Encyclopaedia of seed production of world crops*. Chichester, UK: John Wiley and Sons.

Kobayashi, M., J.-Z. Lin, J. Davis, L. Francis, and M. T. Clegg. 2000.

Quantitative analysis of avocado outcrossing and yield in California using RAPD markers. *Scientia Horticulturae* 86:135–49.

Koeppel, D. 2008. *Banana: The fate of the fruits that changed the world.* New York: Plume.

Larsen, K. 1977. Self-incompatibility in *Beta vulgaris* L. I. Four gametophytic, complementary S-loci in sugar beet. *Heredity* 85:227–48.

Laughlin K. D., A. G. Power, A. A. Snow, and L. J. Spencer. 2009. Risk assessment of genetically engineered crops: Fitness effects of virus-resistance transgenes in wild *Cucurbita pepo. Ecological Applications* 19:1091–101.

Lively, C. M., and L. T. Morran. 2014. The ecology of sexual reproduction. *Journal of Evolutionary Biology* 27:1292–303.

Lloyd, D. G., and C. J. Webb. 1986. The avoidance of interference between the presentation of pollen and stigmas in angiosperms I. Dichogamy. *New Zealand Journal of Botany* 24:165–82.

Longden, P. C. 1989. Effect of increasing weed-beet density on sugarbeet yield and quality. *Annals of Applied Biology* 114:527–32.

———. 1993. Weed beet: A review. *Aspects of Applied Biology* 35:185–94.

Martineau, B. 2001. *First fruit: The creation of the Flavr Savr tomato and the birth of genetically engineered food.* New York: McGraw Hill.

Mayer R. S., and M. D. Purugganan. 2013. Evolution of crop species: Genetics of domestication and diversification. *Nature Reviews Genetics* 14: 840–52.

Maynard Smith, J. 1971. What use is sex? *Journal of Theoretical Biology* 30:319–35.

Mazetti, K. 2008. *Benny and Shrimp*. Translation from the Swedish by Sarah Death. London: Penguin.

McGee, H. 2004. *On food and cooking: The science and lore of the kitchen*. Rev. ed. New York: Scribner.

Michod, R. E. 1997. What good is sex? *The Sciences* 37:42–46.

Molina, R. T., A. M. Rodríguez, I. S. Palaciso, and F. G. López. 1996. Pollen production in anemophilous trees. *Grana* 35:38–46.

Mücher, T., P. Hesse, M. Pohl-Orf, N. C. Ellstrand, and D. Bartsch. 2000. Characterization of weed beet in Germany and Italy. *Journal of Sugar Beet Research* 37(3):19–38.

Nakajima, N., and Y. Matsuura. 1997. Purification and characterization of konjac glucomannan degrading enzyme from anaerobic human intestinal bacterium, *Clostridium butyricum-Clostridium beijerinckii* group. *Bioscience, Biotechnology, and Biochemistry* 61:1739–42.

National Academies of Sciences, Engineering, and Medicine. 2016. *Genetically Engineered Crops: Experiences and Prospects.* Washington, DC: National Academies Press.

National Research Council. 2002. *Environmental effects of transgenic plants.* Washington, DC: National Academy Press.

———. 2004. *Biological confinement of genetically engineered organisms.* Washington, DC: National Academy Press.

Newman, S. E., and A. S. O'Connor. 2009. *Edible flowers.* Colorado State University Extension factsheet. 7.237.

Ordonez, N., M. F. Seidl, C. Waalwijk, A. Drenth, A. Kilian, B. P. H. J. Thomma, R. C. Ploetz et al. 2015. Worse comes to worst: Bananas and Panama disease—When plant and pathogen clones meet. *PLoS Pathogens* 11: e1005197. doi:10.1371/journal.ppat.100519.

Otto, S. P., and A. C. Gerstein. 2006. Why have sex? The population genetics of sex and recombination. *Biochemical Society Transactions* 34:519–22.

Owen, F. V. 1942. Inheritance of cross- and self-sterility and self-fertility in *Beta vulgaris. Journal of Agricultural Research* 69: 679–98.

Owen, M. D. K. 2005. Maize and soybeans—controllable volunteerism without ferality? In *Crop ferality and volunteerism*, ed. J. Gressel, pp. 209–30. Boca Raton, FL: CRC Press.

Peterson, P. A. 1955. Avocado flower pollination and fruit set. *California Avocado Society 1955 Yearbook* 39:163–69.

Pineau, C., and M. Westgate. 2013. *The non-GMO cookbook: Recipes and advice for a non-GMO lifestyle.* New York: Skyhorse Publishing.

Pollan, M. 2001. *The botany of desire: A plant's eye-view of the world.* New York: Random House.

———. 2006. *The omnivore's dilemma: A natural history of four meals.* New York: Penguin.

Prendeville, H. R., X. Ye, T. J. Morris, and D. Pilson. 2012. Virus infections in wild plant populations are both frequent and often unapparent. *American Journal of Botany* 99:1033–42.

Quemada, H., L. Strehlow, D. S. Decker-Walters, and J. E. Staub. 2008. Population size and incidence of virus infection in free-living populations of *Cucurbita pepo*. *Environmental Biosafety Research* 7:185–96.

Reichman, J. R., L. S. Watrud, E. H. Lee, C. A. Burdick, M. A. Bollman, M. A. Storm, G. A. King et al. 2006. Establishment of transgenic herbicide-resistant creeping bentgrass (*Agrostis stolonifera* L.) in nonagronomic habitats. *Molecular Ecology* 15:4243–55.

Renner, S. S. 2014. The relative and absolute frequencies of angiosperm sexual systems: Dioecy, gynodioecy, monoecy, and an updated online database. *American Journal of Botany* 101:1588–96.

Richards, A. J. 1997. *Plant breeding systems*. 2nd ed. London: Chapman and Hall.

Rick, C. M. 1978. The tomato. *Scientific American* 239(2):76–87.

———. 1988. Evolution of mating systems in cultivated plants. In *Plant evolutionary biology*, ed. L. Gottlieb and S. Jain, 133–47. London: Chapman and Hall.

———. 1995. Tomato. In *Evolution of crop plants*, 2nd ed., ed. J. Smartt and N. W. Simmonds, pp. 452–57. Harlow, UK: Longman.

Rissler, J., and M. Mellon. 1996. *The ecological risks of engineered crops*. Cambridge, MA: MIT Press.

Ristaino, J. B. 2002. Tracking historic migrations of the Irish potato famine pathogen, *Phytophthora infestans*. *Microbes and Infection* 4:1369–77.

Roach, B. T. 1995. In *Evolution of crop plants*, 2nd ed., ed. J. Smartt and N. W. Simmonds, pp. 160–65. Harlow, UK: Longman.

Robinson, J. C. 1996. *Bananas and plantains*. Cambridge: CAB International.

Ronald, P. M., and R. W. Adamchak. 2008. *Tomorrow's table: Organic farming, genetics, and the future of food*. New York: Oxford University Press.

Salazar-García, S., L. C. Garner, and C. J. Lovatt. 2013. Reproductive biology. In *The avocado: Botany, production, and uses*, 2nd ed., ed. B. Schaffer, B. N. Wolstenholme, and A. W. Whiley, pp. 118–67. Wallingford, UK: CABI Publishing.

Santoni, S., and A. Bervillé. 1992. Evidence for gene exchanges between sugar beet (*Beta vulgaris* L.) and wild beets: Consequences for transgenic sugar beets. *Plant Molecular Biology* 20:578–80.

Sasu, M. A., M. J. Ferrari, D. Du, J. A. Winsor, and A. G. Stephenson. 2009. Indirect costs of a nontarget pathogen mitigate the direct benefits of a virus-resistant transgene in wild *Cucurbita*. *Proceedings of the National Academy of Sciences of the United States of America* 106:19067–71.

Schaffer, B., B. N. Wolstenholme, and A. W. Whiley. 2013. Introduction to *The avocado: Botany, production, and uses*, 2nd ed., ed. B. Schaffer, B. N. Wolstenholme, and A. W. Whiley, pp. 1–9. Wallingford, UK: CABI Publishing.

Sedgley, M. 1979. Inter-varietal pollen tube-growth and ovule penetration in the avocado. *Euphytica* 28:25–35.

Sharples, B. H. 1919. When is an avocado ripe? How to tell a ripe fruit. *California Avocado Association Annual Report*, 30–31.

Sherkow, J. S., and H. T. Greely. 2013. What if extinction is not forever? *Science* 340:33–34.

Shepherd, J., and G. Bender. 2002. A history of the avocado industry in California. *California Avocado Society Yearbook* 85:29–50.

Simmonds, N. W. 1966. *Bananas*. 2nd ed. London: Longmans.

———. 1979. *Principles of crop improvement*. London: Longmans.

———. 1995. Bananas. In *Evolution of crop plants*, 2nd ed., ed. J. Smartt and N. W. Simmonds, pp. 370–75. Harlow, UK: Longmans.

Singh, R. P., D. R. Hodson. J. Huerta-Espino, Y. Jin, S. Bhavani, P. Njau, S. Herrera-Foessel et al. 2011. The emergence of Ug99 races of the stem rust fungus is a threat to world wheat production. *Annual Review of Phytopathology* 49:465–81.

Sites, J. W., D. M. Peccinini-Seale, C. Moritz, J. W. Wright, and W. M. Brown. 1990. The evolutionary history of parthenogenetic *Cnemidophorus lemniscatus* (Sauria, Teiidae). I. Evidence for a hybrid origin. *Evolution* 44:906–21.

Silverman, M. 1977. *A city herbal*. New York: Knopf.

Smith, A. F. 2013. Sugar: A global history. London: Reaktion Books.

Snow, A. A. 2012. Illegal gene flow from transgenic creeping bentgrass: The saga continues. *Molecular Ecology* 21: 4663–64.

Soukup, J., and J. Holec. 2004. Crop-wild interaction within the *Beta vulgaris* complex: Agronomic aspects of weed beet in the Czech Republic. In *Introgression from genetically modified plants into wild relatives*, ed. H. C. M. den Nijs, D. Bartsch, and J. Sweet, pp. 203–18. Wallingford, UK: CABI Publishing.

Stace, C. A. 1975. *Hybridization and the flora of the British Isles.* London: Academic Press.

Standage, T. 2005. *A history of the world in 6 glasses.* New York: Walker and Co.

Stout, A. B. 1923. A study in cross-pollination of avocados in Southern California. *California Avocado Association Annual Report* 7:29–45.

Stover, R. H., and N. W. Simmonds. 1987. *Bananas*, Tropical agricultural series, 3rd ed. London: Longmans.

Suojala, T. 2000. Variation in sugar content and composition of carrot storage roots at harvest and during storage. *Scientia Horticulturae* 85:1–19.

Swift, J. (1726) 1999. *Gulliver's travels.* The Pennsylvania State University. www2.hn.psu.edu/faculty/jmanis/swift/g~travel.pdf.

Syvanen, M., and C. I. Kado. 2012. *Horizontal gene transfer.* 2nd ed. San Diego, CA: Academic Press.

Taylor, P. E., G. Card, J. House, M. H. Dickinson, and R. C. Flagna. 2006. High-speed pollen release in the white mulberry tree, *Morus alba* L. *Theoretical and Applied Genetics* 19:19–24.

Thangavelu, R., and M. M. Mustaffa. 2010. First report on the occurrence of a virulent strain of Fusarium wilt pathogen (Race-1) infecting Cavendish (AAA) group of bananas in India. *Plant Disease* 94:1379.

Tomar, N. S., A. Goel, M. Mehra, S. Majumdar, S. D. Kharche, S. Bag, D. Malakar et al. 2015. Difference in chromosomal pattern and relative expression of development and sex related genes in parthenogenetic vis-a-vis fertilized turkey embryos. *Journal of Veterinary Science and Technology* 6:226. doi:10.4172/2157-7579.1000226.

Turkington, R. 2010. Obituary: John L. Harper FRS, CBE 1925–2009. *Bulletin of the Ecological Society of America* 91:9–13.

United Nations. 2014. *World urbanization prospects: The 2014 revision highlights*. Department of Economic and Social Affairs. Population Division, United Nations.

USDA-APHIS. 2017. Animal Plant Health and Inspection Service (APHIS). "Biotechnology (BRS): Permits, Notifications, and Petitions." https://www.aphis.usda.gov/aphis/ourfocus/biotechnology/permits-notifications-petitions/sa_permits/ct_status/.

USDA-FAS. 2016. *Sugar: World markets and trade.* United States Department of Agriculture, Foreign Agricultural Service, Office of Global Analysis..

Van Valen, L. 1973. A new evolutionary law. *Evolutionary Theory* 1: l–30.

Viard, F., J. Bernard, and B. Desplanque. 2002. Crop-weed interactions in the *Beta vulgaris* complex at a local scale: Allelic diversity and gene flow within sugar beet fields. *Theoretical and Applied Genetics* 104:688–97.

Vrecenar-Gadus, M., and N. C. Ellstrand. 1985. The effect of planting design on outcrossing rate and yield in the 'Hass' avocado. *Scientia Horticulturae* 27:215–21.

Waites, G. E. II., C. Wang, and P. D. Griffin. 1998. Gossypol: Reasons for its failure to be accepted as a safe, reversible male antifertility drug. *International Journal of Andrology* 21:8–12.

Waltz, E. 2012. Tiptoeing around transgenics. *Nature Biotechnology* 30:215–17.

Warring, A. 2013. Counting their blessings and giving back. *Citrograph* 4(1):22–30.

Warschefsky, E. J., L. L. Klein, M. H. Frank, D. H. Chitwood, J. P. Lando, E. J. B. von Wettberg, and A. J. Miller. 2016. Rootstocks: Diversity, domestication, and impacts on shoot phenotypes. *Trends in Plant Science* 21:418–37.

Warwick, S. I., A. Légère, M.-J. Simard, and T. J. James. 2008. Do escaped transgenes persist in nature? The case of an herbicide resistance transgene in a weedy *Brassica rapa* population. *Molecular Ecology* 17:1387–95.

Watrud, L. S., E. H. Lee, A. Fairbrother, C. Burdick, J. R. Reichman, M. Bollman, M. Storm et al. 2004. Evidence for landscape-level, pollen-mediated gene flow from genetically modified creeping bentgrass with CP4 EPSPS as a marker. *Proceedings of the National Academy of Sciences of the United States of America* 101: 14533–38.

Wendel, J. F. 1995. Cotton. In *Evolution of crop plants*, 2nd ed., ed. J. Smartt and N. W. Simmonds, pp. 358–66. Harlow, UK: Longman.

Whiley, A. W., B. N. Wolstenholme, and G. S. Bender. 2013. Crop management. In *The avocado: Botany, production, and uses*, 2nd ed., ed. B. Schaffer, B. N. Wolstenholme, and A. W. Whiley, pp. 342–79. Wallingford, UK: CABI Publishing.

Williams, G. C. 1975. *Sex and evolution*. Princeton, NJ: Princeton University Press.

Wilson, H. D. 1993. Free-living Cucurbita pepo in the United States: Viral resistance, gene flow, and risk assessment. Prepared for USDA Animal and Plant Health Inspection. Hyattsville, MD: USDA-APHIS.

Wipff, J. K., and C. Fricker. 2001. Gene flow from transgenic creeping bentgrass (*Agrostis stolonifera* L.) in the Willamette Valley, Oregon. *International Turfgrass Society Research Journal* 9:224–42.

Wuethrich, B. 1998. Why sex? Putting theory to the test. *Science* 281:1980–82.

Wutscher, H. K. 1979. Citrus rootstocks. *Horticultural Reviews* 1: 237–69.

Zapiola, M. L., C. K. Campbell, M. D. Butler, and C. A. Mallory-Smith. 2008. Escape and establishment of transgenic glyphosate-resistant creeping bentgrass *Agrostis* stolonifera in Oregon, USA: A 4-year study. *Journal of Applied Ecology* 45:486–94.

Zapiola, M. L., and C. A. Mallory-Smith. 2017. Pollen-mediated gene flow from transgenic perennial creeping bentgrass and hybridization at the landscape level. *PLoS ONE* 12(3):e0173308. doi:10.1371/journal.pone.0173308.

Zuckerman, C. 2013. Meet the Lumper: Ireland's old new potato. *National Geographic News*. http://news.nationalgeographic.com/news/2013/03/130315-irish-famine-potato-lumper-food-science-culture-ireland/.

Index

Note: page numbers in *italics* refer to illustrations.